SpringerBriefs in Mathematics

SpringerBriefs in Mathematics showcases expositions in all areas of mathematics and applied mathematics. Manuscripts presenting new results or a single new result in a classical field, new field, or an emerging topic, applications, or bridges between new results and already published works, are encouraged. The series is intended for mathematicians and applied mathematicians.

For further volumes:
http://www.springer.com/series/10030

Tuğrul Dayar

Analyzing Markov Chains using Kronecker Products

Theory and Applications

 Springer

Tuğrul Dayar
Department of Computer Engineering
Bilkent University
Ankara, Turkey

ISSN 2191-8198   ISSN 2191-8201 (electronic)
ISBN 978-1-4614-4189-2   ISBN 978-1-4614-4190-8 (eBook)
DOI 10.1007/978-1-4614-4190-8
Springer New York Heidelberg Dordrecht London

Library of Congress Control Number: 2012939483

Mathematics Subject Classification (2010): 60J27, 60J10, 60-04, 15A69, 15B51, 15-04, 65C20, 65C40, 65F50, 65F10, 65F08, 65F05, 65-04, 37A30, 37A50, 37B25, 93D30, 37-04, 93-04, 60J28, 60J20, 68M15, 90B25, 68M20, 68-04, 90B22, 90-04, 80A30, 80-04, 92E20, 92C40, 92C42, 92C45, 92D25, 92-04

Printed on acid-free paper

Springer is part of Springer Science+Business Media (www.springer.com)

To the memory of my father

Preface

This book has grown out of work spanning the last 15 years. Billy Stewart and Brigitte Plateau introduced me to stochastic automata networks (SANs) during my visit to Grenoble in June 1996. A study on robotic tape libraries using SANs was carried out in August that year with Odysseas Pentakalos and Brooke Stephens in Greenbelt, Maryland. The help received during this process from Paulo Fernandes regarding the software was instrumental. Later I had many interesting talks with Jean-Michele Fourneau and Franck Quessette over SANs in Versailles and Ankara from 1998 to 2000. Nihal Pekergin was also present during these visits and brought her expertise on stochastic comparison into the picture. In June 2000, we had enjoyable discussions with Ivo Marek and Petr Mayer in Prague on the convergence properties of iterative aggregation–disaggregation. In the academic year 2002–2003, I got a chance to learn about hierarchical Markovian models (HMMs) from Peter Buchholz in Dresden. Our discussions on HMMs continued in Dortmund in June 2005 and at Dagstuhl in February 2007. It was in Dortmund and then Ankara where we investigated compositional Markovian models for symmetries with Peter Kemper in 2005. Kishor Trivedi visited Ankara in 1997; he was always available by e-mail in 2008 while we were writing a joint paper and was ready for further discussions later that year in Seattle. In early 2010 in Saarbrücken, I was convinced by Verena Wolf of the difficulties associated with analyzing systems of stochastic chemical kinetics using SANs and HMMs. This also led to stimulating exchanges with Holger Hermanns, Werner Sandmann, and David Spieler. I thank them all for providing a scholarly atmosphere in which to carry out research and for their time. I am grateful to Bilkent University for being very understanding and generous in granting these research leaves without which it would not have been possible to write this book. I am also fortunate to have worked with a number of students at Bilkent who found compositional Kronecker models for Markovian systems such as SANs and HMMs interesting and exciting: Ertuğrul Uysal, Oleg Gusak, Akın Meriç, İlker Nadi Bozkurt, and Muhsin Can Orhan. Grants available in one form or another over the years allowed me to continue working on the subject from the Turkish Scientific and Technological Research Council, the French National Scientific Research Center, the Center of Excellence in Space Data and

Information Systems, the Czech Technical University, the Alexander von Humbodt Foundation, the Turkish Academy of Sciences, and the Cluster of Excellence in Multimodal Computing and Interaction. Finally, I thank Donna Chernyk and her team at Springer, New York for initiating the book project, walking me through the process, and making it happen. I also thank Lesley Poliner from Springer for the editorial process she conducted and Udaiyar Rekha from SPi Technologies, India for her effort during the production of the book. Parts of the book appeared in the proceedings of the Markov Anniversary Meeting in 2006 [55]. I hope you enjoy the outcome.

Ankara, Turkey   Tuğrul Dayar

Contents

Chapter 1
Introduction

This book is about the numerical analysis of very large discrete event dynamic systems, including those with countably infinite state spaces, satisfying Markovian assumptions. It reviews the progress made in the last 25 years regarding the probabilistic analysis of such systems represented compactly using Kronecker products of smaller subsystem transition matrices. Although the Kronecker representation does not provide a solution to the storage problem of state probability vectors associated with the system, it enables the storage of the underlying state transition matrix compactly, thereby facilitating the analysis of multidimensional systems that are an order of magnitude larger than those that can be analyzed with conventional sparse matrix techniques on the same platform due to memory limitations.

The subject matter is interdisciplinary and at the intersection of applied mathematics, specifically numerical linear algebra and computational probability, and computer science. Its understanding requires basic linear algebra, probability, discrete mathematics, and high-level programming knowledge. The exposition is concise and rigorous, yet it tries to be complete and touches almost all relevant aspects without being too technical. Since Kronecker notation necessitates the introduction of many indices on variables and is not easy to follow, the reader is walked through various examples in the text to make the discussion more accessible and concrete. The examples come from areas as diverse as systems availability, closed queueing networks, and systems of stochastic chemical kinetics. The last of these also requires the use of concepts from stability theory to aid in truncating the respective countably infinite state spaces judiciously for analysis purposes. Markov population models fall into this class and can be analyzed similarly. Up until recently, stochastic simulation seemed to be the only viable approach that would yield relatively accurate results for this class of problems. With the existing continuous improvement in computer technology, it only makes sense to invest in state-based numerical analysis approaches for very large Markovian systems such as those discussed here to obtain more accurate results at a finer level of detail.

Even though the book is not specifically about modeling using Kronecker products, it mentions a sufficient number of modeling formalisms, comparing

T. Dayar, *Analyzing Markov Chains using Kronecker Products: Theory and Applications*, SpringerBriefs in Mathematics, DOI 10.1007/978-1-4614-4190-8_1, © Tuğrul Dayar 2012

and contrasting them along the way, so that the reader should have an idea as to how to proceed after having finished reading the book. Matlab codes used in modeling and analyzing problems from closed queueing networks and systems of stochastic chemical kinetics using Kronecker products are made publicly available. When space is not sufficient to discuss a particular issue in detail, references to the literature are provided for further reading. Open research areas are also indicated throughout the text as they become pertinent. Having set the stage with our motivation to gather all this information in one place and write the book, we can now start our discussion.

We consider discrete state space Markovian processes called *Markov chains* (MCs). These are stochastic processes that start in a particular state, evolve by visiting states in its state space using the available transitions, and exhibit the *memoryless* property. This property dictates that the probability distribution of the next state of a process conditioned on the current state and the previous states the process has visited depends only on the current state. In *discrete time*, this requires that the time spent in a state during a visit be *geometrically* distributed, whereas in *continuous time*, it requires that the time spent in a state during a visit be *exponentially* distributed.

Throughout this book, we mainly consider *continuous-time MCs* (CTMCs) and briefly touch on *discrete-time MCs* (DTMCs) wherever appropriate. We adhere to the convention that probability vectors are row vectors; otherwise, all vectors are column vectors as in linear algebra. $\mathbf{e}$ represents a column vector of 1s and $\mathbf{e}_i$ represents the ith principal axis vector, that is, the ith column of the identity matrix, I. We use calligraphic letters for sets, uppercase letters for matrices, and lowercase letters for vectors. The Greek letter $\boldsymbol{\pi}$ is used to denote a particular vector. Otherwise, we use Greek letters to denote problem-specific real-valued parameters. We indicate entries of matrices using lowercase letters and the row/column indices in parentheses. Similarly, an entry of a vector is shown by writing the corresponding index in parentheses. We start indices from zero, by which it is possible to represent an empty (sub)system. An exception is state vectors, whose components are state variables indicated by subscripts starting from one. diag($\mathbf{a}$), subdiag($\mathbf{a}$), and superdiag($\mathbf{a}$) denote matrices with entries of vector $\boldsymbol{a}$ along their diagonal, subdiagonal, and superdiagonal, respectively; all other entries of these matrices are zero. We use a subscript under I to indicate its order. Similarly, the subscript $m \times n$ under a matrix indicates that the matrix is ($m \times n$). The lengths of the vectors are determined by the context in which they are used. $\mathbb{R}$ and $\mathbb{Z}$ stand respectively for the set of real numbers and integers, whereas $\mathbb{R}_{\geq 0}$ and $\mathbb{Z}_+$ denote their nonnegative subsets.

For the time being, we assume that the state space of the CTMC is finite but later relax this assumption when we discuss infinite level-dependent quasi-birth-and-death (LDQBD) processes. Thus, now let the cardinality of the state space be n. A CTMC with n states may be represented by an ($n \times n$) matrix $Q \in \mathbb{R}^{n \times n}$. This matrix is also known as the *infinitesimal generator* (or *transition rate*) matrix of the associated Markovian process [81, 139]. The off-diagonal entries of this matrix are nonnegative and indicate the rates of exponentially distributed transition times between pairs of states. In other words, $q(i, j)$ for $i \neq j$ denotes the exponential rate of time by which the process makes a transition from state i to state j. The diagonal of Q is formed by the negated row sums of its off-diagonal entries (i.e., $q(i, i) = -\sum_{j \neq i} q(i, j)$) and, hence, is nonpositive. Throughout our discussion, we assume that entries of Q are independent of time, that is, Q is *time-homogeneous*. Treatment of time-inhomogeneous CTMCs is beyond the scope of this book.

Example 1. Here is an infinitesimal generator matrix corresponding to a CTMC with 12 states:

$$
Q =
\begin{array}{c}
\\ 0 \\ 1 \\ 2 \\ 3 \\ 4 \\ 5 \\ 6 \\ 7 \\ 8 \\ 9 \\ 10 \\ 11
\end{array}
\begin{array}{c}
\begin{array}{cccccccccccc}
0 & 1 & 2 & 3 & 4 & 5 & 6 & 7 & 8 & 9 & 10 & 11
\end{array} \\
\left(
\begin{array}{cccccccccccc}
* & \lambda_3 & \lambda_2 & & & & \lambda_1 & & & & & \\
\mu_3 & * & & \lambda_2 & & & & \lambda_1 & & & & \\
\mu_2 & & * & \lambda_3 & \lambda_2 & & & & \lambda_1 & & & \\
& \mu_2 & \mu_3 & * & & \lambda_2 & & & & \lambda_1 & & \\
& & \mu_2 & & * & \lambda_3 & & & & & \lambda_1 & \\
& & & \mu_2 & \mu_3 & * & & & & & & \lambda_1 \\
\mu_1 & & & & & & * & \lambda_3 & \lambda_2 & & & \\
& \mu_1 & & & & & \mu_3 & * & & \lambda_2 & & \\
& & \mu_1 & & & & \mu_2 & & * & \lambda_3 & \lambda_2 & \\
& & & \mu_1 & & & & & \mu_2 & \mu_3 & * & \lambda_2 \\
& & & & \mu_1 & & & & & \mu_2 & * & \lambda_3 \\
\mu & & & & & \mu_1 & & & & & \mu_2 & \mu_3 & *
\end{array}
\right),
\end{array}
\tag{1.1}
$$

where $*$ denotes the negated off-diagonal row sums. Note that Q has 41 nonzero off-diagonal entries among 132 possible ones; hence, it would normally be classified as a *sparse* matrix [64, 130, 139, 143].

It is well known that the minimum of exponentially distributed random variables is exponentially distributed with rate parameter equal to the sum of the rate parameters of the random variables. Therefore, the absolute value of a diagonal entry may be perceived as the rate of the exponentially distributed exit time from that particular state. This implies that row sums of Q are necessarily zero.

Let the *initial* probability distribution vector of Q be denoted by π_0. Clearly, $\pi_0 \in \mathbb{R}^{1 \times n}$, $\pi_0 \geq 0$, and $\pi_0 e = 1$. The ith entry of π_0, written $\pi_0(i)$, denotes the probability of the process being in state i at time 0. Then the *transient* probability distribution vector at time $t \geq 0$ is given by

$$
\pi_t = \pi_0 e^{Qt} = \pi_0 e^{-\Gamma t} \sum_{k=0}^{\infty} \frac{(\Gamma t)^k}{k!} \left(I + \frac{1}{\Gamma} Q \right)^k,
\tag{1.2}
$$

where $\Gamma = \max_i |q_{i,i}|$ [80, 84]. This vector gives the probability distribution at time t when the process starts according to π_0 at time 0. Equation (1.2) is the simplest way for the transient vector to be stably computed and is known by the name of *uniformization* (or *randomization, Jensen's method*). In passing to steady-state analysis, we remark that almost all methods for transient analysis are based on vector–matrix multiplications and will not be discussed here further.

There are processes for which $\lim_{t \to \infty} \pi_t$ exists. Whenever this limit exists, we define $\pi := \lim_{t \to \infty} \pi_t$ and refer to π as the *steady-state* (or *limiting, long-run*) probability distribution vector. The steady-state vector satisfies

$$\pi Q = 0, \quad \pi e = 1, \tag{1.3}$$

and, hence, is also a *stationary* distribution [81, 139]. For finite CTMCs, the necessary and sufficient condition for the existence of π is *irreducibility*, that is, each state must be reachable from every other state by following the nonzero transitions in Q. Many properties of transition matrices underlying finite MCs can be found in [13, 93, 135]. Conditions for the existence of steady state in irreducible CTMCs with countably infinite state spaces are given in Chap. 5.

In the Kronecker-based approach, Q is represented using Kronecker products [50, 148] of smaller matrices and is never explicitly generated. This is motivated by the fact that systems are usually comprised of interacting subsystems and, hence, have multidimensional state spaces. Each subsystem can be considered as representing a separate dimension in this multidimensional state space. The state of a subsystem may either evolve independently, that is, in isolation from other subsystems, or it may evolve in synchronization with other subsystems. To the former case there corresponds a Kronecker product with all factors, except one associated with the particular subsystem, being identity matrices. To the latter case there corresponds a Kronecker product with all factors, except those associated with the subsystems that need to be synchronized, being identity matrices. Hence, Q can be represented using sums of such Kronecker products. The implementation of transient solvers for (1.2) and steady-state solvers for (1.3) can rest on this compact Kronecker representation thanks to the existence of an efficient vector–Kronecker product multiplication algorithm known as the *shuffle algorithm* [50].

Example 1. (ctnd.) The CTMC corresponding to (1.1) is the model of a system [4] with three subsystems having respectively one, two, and one redundant component(s), only one of which works in each subsystem. The working component in each subsystem fails independently of other subsystems at rates of λ_1, λ_2, and λ_3, respectively. When a working component fails, it is replaced immediately with an intact redundant component (if there is any left) in the same subsystem. Furthermore, there is one repairman in each subsystem who can repair a failed component independently of other subsystems at rates of μ_1, μ_2, and μ_3, respectively. When all components fail, the system is brought up by a global repair at rate μ. To give some examples, in (1.1) state 0 corresponds to all subsystems being intact, state 11 corresponds to all subsystems having failed, and state 6 corresponds to a failed first subsystem while the second and third subsystems are intact. Overall, this is a system that tries to improve steady-state availability using a larger number of redundant components, which themselves do not need to be highly reliable.

In practice, the representation of Q based on Kronecker products is obtained using various modeling formalisms. These include compositional Markovian models such as stochastic automata networks (SANs) [119–122, 139] and different classes of superposed stochastic Petri nets (SPNs) [63, 94], hierarchical Markovian models (HMMs) of queueing networks [19], generalized stochastic Petri nets (GSPNs) [36], or systems of asynchronously communicating stochastic modules [39], and stochastic process algebras such as the performance evaluation process algebra (PEPA) [91]. These modeling formalisms are integrated with various software packages such as the Abstract Petri Net Notation (APNN) toolbox [3, 7], the Performance Evaluation of Parallel Systems (PEPS) software tool [8, 15, 118] (see also GTAexpress [48]), the PEPA Workbench [45, 117], and the stochastic model-checking analyzer for reliability and timing (SMART) [44, 136].

An advantage of HMMs is their ability to represent Q using Kronecker products without introducing unreachable states. Matrix diagrams [43] and representations for specific models as in [89] can also be used to achieve the same effect when state spaces are expressed compositionally. Other approaches can be used to deal with unreachable states, as discussed in [10, 24, 38]. At the beginning of our discussion, we make the assumption that the product state space of a system composed of subsystems does not have unreachable states and is irreducible. Later we relax this assumption and show how this problem can be tackled using the HMM approach. Yet in many practical applications, Q is very large and has many nonzeros, necessitating its storage in memory using Kronecker products. To efficiently analyze Markovian models based on Kronecker products, various algorithms for vector–Kronecker product multiplication based on the shuffle algorithm are devised [10, 11, 38, 47, 66, 67, 121, 124] and used as kernels in iterative solution techniques proposed for different modeling formalisms. The transient distribution in (1.2) can be computed using vector–Kronecker product multiplications as in [19]. The steady-state distribution in (1.3) also needs to be computed using vector–Kronecker product multiplications since *direct methods* based on complete factorizations, such as *Gaussian elimination* (GE), normally introduce new nonzeros that cannot be accommodated.

The two papers [25] and [141] provide good overviews of iterative solution techniques for the analysis of MCs based on Kronecker products. The solution of lower- and upper-triangular systems of equations arising from a complete factorization of a Kronecker product is the subject of [65]. The approach relies on factorizing the smaller factors making up the Kronecker product. The problem with this approach is that even if the factorization existed, it is not clear why the resulting smaller lower- and upper-triangular factors should also be sparse. Issues related to reachability analysis, vector–Kronecker product multiplication, hierarchical state space generation in Kronecker-based matrix representations for large Markovian models are surveyed in [37]. A comparison of the merits of the SAN and GSPN modeling formalisms using the PEPS and SMART software packages can be found in [42]. Along a different line, [49, 70, 75] give conditions for SANs to have product-form steady-state distributions, whereas transient availability of a grid of 5,000

CPUs is investigated in [16] using SANs. Finally, graph-theoretic arguments are used in [72] to aid the steady-state analysis of SANs.

Although Kronecker representations for CTMCs underlying many models of practical applications have been considered, so far only a handful of DTMCs based on Kronecker products have appeared in the literature. For instance, the one in [123] is a model of synchronization via messages passing in a distributed system, that in [120] is a model of the mutual exclusion algorithm of Lamport, those in [73,74] are models of buffer admission mechanisms for asynchronous transfer mode (ATM) networks from telecommunications, that in [149] is a multiservice resource allocation policy for wireless ATM networks, and that in [69] is a model to compute the loss rate in a multistage ATM switch. The model in [88] is a larger, scalable, and extended version of that in [149]. It serves as a good example showing that the underlying MC of a discrete-time model based on Kronecker products can be relatively dense and numerically difficult to analyze. These case studies are based on the SAN modeling formalism, whereas [134] extends the Kronecker representation to SPNs with discrete phase-type distributions. Sufficient conditions for DTMCs with transition probability matrices in the form of generalized Kronecker products to have product-form solutions are investigated in [71]. Clearly, the area of DTMCs based on Kronecker products can use other case studies and formalisms.

Here, we take a vector–matrix approach and discuss recent results related to the analysis of MCs based on Kronecker products independently of modeling formalisms. In Chap. 2, we provide background material on the Kronecker representation of a CTMC, show that it has a rich structure that is nested and recursive, express the small CTMC in Chap. 1 as a sum of Kronecker products, present the vector–Kronecker product multiplication algorithm, and consider preprocessing of the Kronecker representation so as to expedite numerical analysis. We discuss permuting the nonzero structure of the underlying CTMC symmetrically by reordering, changing the orders of the nested blocks by grouping, and reducing the size of the state space by lumping. Chapter 3 is devoted to the steady-state analysis of CTMCs based on Kronecker products with block iterative methods, preconditioned projection methods, and multilevel methods. For all these iterative methods it is essential to consider splitting the smaller matrices corresponding to subsystems, which we do at the beginning of the chapter. The example introduced in Chap. 1 is again used to illustrate various concepts. In Chap. 4, we recall a decompositional iterative method that is geared toward systems with weakly interacting subsystems. We show that for certain values of its parameters, the example in Chap. 1 forms such a system, and we are better off using the decompositional iterative method rather than the methods in Chap. 3. Then we discuss two approximative decompositional iterative methods for the steady-state analysis of closed queueing networks with phase-type service distributions and arbitrary buffer sizes. This chapter also serves to introduce a method for handling the reachability problem with the help of the HMM approach. Chapter 5 investigates the steady-state analysis of systems of stochastic chemical kinetics modeled as infinite LDQBD processes. It shows how the countably infinite state space can be partitioned into levels and truncated judiciously using Lyapunov functions for steady-state analysis. We conclude the book with Chap. 6. The results can be extended to DTMCs based on Kronecker products with some modifications. Areas that need further research are mentioned within the sections as they are discussed. Finally, we remark that parallel implementations exploiting the Kronecker representation are beyond the scope of this book and form an open area for research.

Chapter 2
Preliminaries

The *Kronecker* (or *tensor*) *product* [50, 148] of two (rectangular) matrices A and B with $A = [a(i_A, j_A)]$ is

$$A \otimes B = [a(i_A, j_A)B].$$

Or, more formally, given $A \in \mathbb{R}^{n_A \times m_A}$ and $B \in \mathbb{R}^{n_B \times m_B}$, $A \otimes B$ yields the (rectangular) matrix $C \in \mathbb{R}^{n_A n_B \times m_A m_B}$ whose entries satisfy

$$c(i_C, j_C) = a(i_A, j_A)b(i_B, j_B),$$

$$\text{with} \quad i_C = i_A n_B + i_B \quad \text{and} \quad j_C = j_A m_B + j_B$$

for

$$(i_A, j_A) \in \{0, \dots, n_A - 1\} \times \{0, \dots, m_A - 1\},$$

$$(i_B, j_B) \in \{0, \dots, n_B - 1\} \times \{0, \dots, m_B - 1\},$$

where $\times$ is the *Cartesian product* operator. Note that in a two-dimensional representation, the row indices of C are in $\{0, \dots, n_A - 1\} \times \{0, \dots, n_B - 1\}$, whereas its column indices are in $\{0, \dots, m_A - 1\} \times \{0, \dots, m_B - 1\}$. Hence, the ordering of rows and columns of C with respect to this two-dimensional representation is *lexicographical* since

$$c(i_C, j_C) = c((i_A, i_B), (j_A, j_B)) = c(i_A n_B + i_B, j_A m_B + j_B).$$

The Kronecker product is *associative* and defined for more than two matrices. To explain this further for a MC setting, let us consider the Kronecker product of H square matrices as in

$$X = X^{(1)} \otimes X^{(2)} \otimes \cdots \otimes X^{(H)} = \otimes_{h=1}^{H} X^{(h)},$$

T. Dayar, *Analyzing Markov Chains using Kronecker Products: Theory and Applications*, SpringerBriefs in Mathematics, DOI 10.1007/978-1-4614-4190-8_2, © Tuğrul Dayar 2012

where $X^{(h)} \in \mathbb{R}^{n_h \times n_h}$ and row/column indices of $X^{(h)}$ are in $\mathcal{S}^{(h)} = \{0, \ldots, n_h - 1\}$ for $h = 1, \ldots, H$. Therefore, $X \in \mathbb{R}^{n \times n}$, with $n = \prod_{h=1}^{H} n_h$, and the ordered H-dimensional tuples

$$\mathbf{i} = (i_1, \ldots, i_H) \in \times_{h=1}^{H} \mathcal{S}^{(h)} \quad \text{and} \quad \mathbf{j} = (j_1, \ldots, j_H) \in \times_{h=1}^{H} \mathcal{S}^{(h)}$$

may be used to represent the row and column indices of X, respectively. Hence, the Kronecker product of H square matrices implies a one-to-one onto mapping between an H-dimensional state space and a one-dimensional (or *flat*) state space that are lexicographically ordered, and naturally the Kronecker product has been used to define MCs having multidimensional state spaces.

2.1 Kronecker Representation

We assume that the infinitesimal generator of the H-dimensional CTMC at hand is represented as a sum of Kronecker products plus a diagonal matrix. Specifically, we have

$$Q = Q_O + Q_D, \quad Q_O = \sum_{k=1}^{K} \bigotimes_{h=1}^{H} Q_k^{(h)}, \quad Q_D = -\mathrm{diag}(Q_O \mathbf{e}), \qquad (2.1)$$

where Q_O and Q_D correspond respectively to the off-diagonal and diagonal parts of Q, K is the number of Kronecker products (or terms) forming Q_O, H is the number of factors in each Kronecker product, $Q_k^{(h)} \in \mathbb{R}_{\geq 0}^{n_h \times n_h}$ for $k = 1, \ldots, K$ and $h = 1, \ldots, H$. Observe that $Q_O \geq 0$ and $Q_D \leq 0$. If row/column indices of $Q_k^{(h)}$ are in $\mathcal{S}^{(h)} = \{0, \ldots, n_h - 1\}$ for $k = 1, \ldots, K$ and $h = 1, \ldots, H$, then the H-dimensional state space of Q is given by

$$\mathcal{S} = \times_{h=1}^{H} \mathcal{S}^{(h)},$$

the *product state space*. Observe that $|\mathcal{S}| = \prod_{h=1}^{H} |\mathcal{S}^{(h)}| = \prod_{h=1}^{H} n_h = n$. Furthermore, the one-dimensional value of state $\mathbf{i} = (i_1, \ldots, i_H) \in \mathcal{S}$, where $i_h \in \mathcal{S}^{(h)}$ for $h = 1, \ldots, H$, is given by

$$i = \sum_{h=1}^{H} i_h \prod_{l=h+1}^{H} n_l.$$

Oftentimes, when implementing an algorithm, one needs to have a mapping from the one-dimensional state space to the multidimensional state space, and vice versa, since solution vectors are one-dimensional, and appropriate entries of them need to be accessed. Therefore, we will be using the one-dimensional and multidimensional representations of states interchangeably throughout the text.

Each factor in this representation is associated with a different subsystem, and there are H of them. Without loss of generality, we assume that the first H of the K Kronecker product terms model *local* transitions, the kth one being that of subsystem k, whereas the remaining $(K - H)$ Kronecker product terms model *synchronized* transitions. In each of the first K terms, there is only one factor different than the identity matrix; for the kth term it is the kth factor. In each of the remaining $(K - H)$ terms, there are at least two factors different than the identity matrix, and those correspond to subsystems that get synchronized by that particular transition. The existence of an identity matrix in a term implies that the corresponding subsystem does not change its state during that particular transition. Hence, only one subsystem may change its state during a local transition, and at least two subsystems may change their states during a synchronized transition. Note that it is also possible to represent the sum of the first K terms of Kronecker products as a single term of *Kronecker sum* and express the remaining $(K - H)$ terms as a sum of Kronecker products. This has been done for the SAN and HMM formalisms. It is our opinion that this complicates the already cumbersome notation further, and we therefore refrain from doing that here.

One needs space for the diagonal matrix Q_D and the matrices in the Kronecker representation of Q_O, meaning a floating-point vector of length $\prod_{h=1}^{H} n_h$ and at most K (sparse) floating-point matrices of order n_h are stored for $h = 1, \ldots, H$. In the worst case, this amounts to a storage space of $n + \sum_{h=1}^{H} nz_{Q^{(h)}}$ floating-point values, where $nz_{Q^{(h)}}$ is the sum of the number of nonzeros in $Q_k^{(h)}$ for $k = 1, \ldots, K$, compared to nz nonzeros required by the flat representation. Q_D can also be expressed as a sum of Kronecker products:

$$Q_D = -\sum_{k=1}^{K} \bigotimes_{h=1}^{H} \text{diag}\left(Q_k^{(h)} \mathbf{e}\right).$$

However, to enable the efficient implementation of numerical solvers, most of the time Q_D is precomputed and stored explicitly.

Now, each nonzero entry of matrix $Q_k^{(h)}$ in (2.1) is located by its row and column indices, which are members of $\mathcal{S}^{(h)}$. In a more general setting, a nonzero entry of $Q_k^{(h)}$ may be a function of states in state spaces other than $\mathcal{S}^{(h)}$ and, thus, a function of nonlocal states. This phenomenon is a byproduct of the modeling process and is utilized in the SAN modeling formalism. These nonzero entries are referred to as *functional transitions*, and the corresponding Kronecker products are said to be *generalized* [119]. Although it is possible to remove functional transitions from a sum of generalized Kronecker products by introducing new terms [121] or factors [10], functional transitions enable a more compact Kronecker representation with denser factor matrices [42]. We do not consider functional transitions in this discussion but indicate that the results extend to generalized Kronecker products wherever appropriate.

There is a rich structure associated with the Kronecker representation in (2.1). This structure is nested and recursive [32–34, 54, 85, 87, 88, 146]. Let level 0 denote the highest level at which Q is perceived as a single block of order $n = \prod_{h=1}^{H} n_h$. At the next level, which we call level 1, Q is an $(n_1 \times n_1)$ block matrix with blocks of order $\prod_{h=2}^{H} n_h$. At level 2, Q is an $(n_1 n_2 \times n_1 n_2)$ block matrix with blocks of order $\prod_{h=3}^{H} n_h$. Continuing in this manner, at level H, Q is a $(\prod_{h=1}^{H} n_h \times \prod_{h=1}^{H} n_h)$, in other words, $(n \times n)$, block matrix with blocks of order 1. More formally, we have

$$b_l = \begin{cases} 1, & l = 0 \\ n_l^2 b_{l-1}, & l = 1, \ldots, H \end{cases} \quad \text{and} \quad o_l = \begin{cases} n, & l = 0 \\ o_{l-1}/n_l, & l = 1, \ldots, H \end{cases}, \quad (2.2)$$

where b_l and o_l denote the number and order of blocks at level $l = 0, 1, \ldots, H$, respectively. Unrolling the recurrences, we obtain

$$b_l = \prod_{h=1}^{l} n_h^2 \quad \text{and} \quad o_l = \prod_{h=l+1}^{H} n_h.$$

At level l there are $\sqrt{b_l}$ blocks, each of order o_l, along the diagonal of Q. Furthermore, block $((i_1, \ldots, i_l), (j_1, \ldots, j_l))$ of Q at level l is given by

$$Q((i_1, \ldots, i_l), (j_1, \ldots, j_l))$$

$$= \sum_{k=1}^{K} \left(\prod_{h=1}^{l} q_k^{(h)}(i_h, j_h) \right) \left(\bigotimes_{h=l+1}^{H} Q_k^{(h)} \right) + Q_D((i_1, \ldots, i_l), (j_1, \ldots, j_l)) \quad (2.3)$$

$$\text{for} \quad l = 0, \ldots, H,$$

where $Q_D((i_1, \ldots, i_l), (j_1, \ldots, j_l))$ is block $((i_1, \ldots, i_l), (j_1, \ldots, y_l))$ of Q_D with the understanding that $l = 0$ yields Q and $l = H$ yields the scalar

$$q((i_1, \ldots, i_H), (j_1, \ldots, j_H)) = \sum_{k=1}^{K} \prod_{h=1}^{H} q_k^{(h)}(i_h, j_h)$$

$$+ q_D((i_1, \ldots, i_H), (j_1, \ldots, j_H)).$$

Observe that $Q_D((i_1, \ldots, i_l), (j_1, \ldots, j_l)) = 0$ if $(i_1, \ldots, i_l) \neq (j_1, \ldots, j_l)$, meaning it is an off-diagonal block at level l. The nested structure associated with (2.1) is also valid in the presence of functional transitions.

Example 1 (ctnd.). We introduce a small example to illustrate the Kronecker representation associated with a CTMC. Consider the following matrices for a three-dimensional problem, respectively with dimensions of 2, 3, and 2 states, and having four terms of Kronecker products:

$$Q_2^{(1)} = Q_3^{(1)} = I_2, \quad Q_1^{(1)} = \begin{pmatrix} & \lambda_1 \\ \mu_1 & \end{pmatrix}, \quad Q_4^{(1)} = \begin{pmatrix} & \\ & \mu \end{pmatrix},$$

$$Q_1^{(2)} = Q_3^{(2)} = I_3, \quad Q_2^{(2)} = \begin{pmatrix} & \lambda_2 & \\ \mu_2 & & \lambda_2 \\ & \mu_2 & \end{pmatrix}, \quad Q_4^{(2)} = \begin{pmatrix} & & \\ & & \\ & & 1 \end{pmatrix},$$

$$Q_1^{(3)} = Q_2^{(3)} = I_2, \quad Q_3^{(3)} = Q_4^{(3)} = \begin{pmatrix} & \lambda_3 \\ \mu_3 & \end{pmatrix}, \quad Q_4^{(3)} = \begin{pmatrix} & \\ 1 & \end{pmatrix}.$$

Then, from (2.1)

$$Q = \sum_{k=1}^{4} \bigotimes_{h=1}^{3} Q_k^{(h)} + Q_D,$$

and Q is given by (1.1). Now, if we assume that the absence of a matrix in the Kronecker representation indicates that it is identity, then it is possible to do without storing identity matrices. With this understanding, the number of floating-point values stored in the Kronecker representation of Q is 11 for matrices and 12 for diagonal entries, totaling 23, whereas it is 53 for the flat representation in (1.1).

The discrepancy between Kronecker and flat representations becomes substantial for larger values of the state-space size, n, and for denser subsystem matrices. It is also possible to take advantage of identity matrices in the vector–Kronecker product multiplication algorithm discussed in the next section.

Example 1 (ctnd.). Continuing the same example, since $n = 12$ due to $n_1 = n_3 = 2$, and $n_2 = 3$, the recursive definition in (2.2) reveals the nested block structure of Q as follows. At level 0, we have an order 12 matrix [cf. (1.1)]:

$$
\begin{array}{cccccccccccc}
0 & 0 & 0 & 0 & 0 & 0 & 1 & 1 & 1 & 1 & 1 & 1 \\
0 & 0 & 1 & 1 & 2 & 2 & 0 & 0 & 1 & 1 & 2 & 2 \\
0 & 1 & 0 & 1 & 0 & 1 & 0 & 1 & 0 & 1 & 0 & 1
\end{array}
$$

$$
Q = \begin{array}{c}
0\,0\,0 \\
0\,0\,1 \\
0\,1\,0 \\
0\,1\,1 \\
0\,2\,0 \\
0\,2\,1 \\
1\,0\,0 \\
1\,0\,1 \\
1\,1\,0 \\
1\,1\,1 \\
1\,2\,0 \\
1\,2\,1
\end{array}
\left(
\begin{array}{cccccccccccc}
* & \lambda_3 & \lambda_2 & & & & \lambda_1 & & & & & \\
\mu_3 & * & & \lambda_2 & & & & \lambda_1 & & & & \\
\mu_2 & & * & \lambda_3 & \lambda_2 & & & & \lambda_1 & & & \\
& \mu_2 & \mu_3 & * & & \lambda_2 & & & & \lambda_1 & & \\
& & \mu_2 & & * & \lambda_3 & & & & & \lambda_1 & \\
& & & \mu_2 & \mu_3 & * & & & & & & \lambda_1 \\
\mu_1 & & & & & & * & \lambda_3 & \lambda_2 & & & \\
& \mu_1 & & & & & \mu_3 & * & & \lambda_2 & & \\
& & \mu_1 & & & & \mu_2 & & * & \lambda_3 & \lambda_2 & \\
& & & \mu_1 & & & & \mu_2 & \mu_3 & * & & \lambda_2 \\
& & & & \mu_1 & & & & \mu_2 & & * & \lambda_3 \\
\mu & & & & & \mu_1 & & & & \mu_2 & \mu_3 & *
\end{array}
\right) ;
$$

at level 1, we have a (2×2) block matrix with blocks of order 6:

$$
Q = \left(
\begin{array}{cccccc|cccccc}
* & \lambda_3 & \lambda_2 & & & & \lambda_1 & & & & & \\
\mu_3 & * & & \lambda_2 & & & & \lambda_1 & & & & \\
\mu_2 & & * & \lambda_3 & \lambda_2 & & & & \lambda_1 & & & \\
& \mu_2 & \mu_3 & * & & \lambda_2 & & & & \lambda_1 & & \\
& & \mu_2 & & * & \lambda_3 & & & & & \lambda_1 & \\
& & & \mu_2 & \mu_3 & * & & & & & & \lambda_1 \\
\hline
\mu_1 & & & & & & * & \lambda_3 & \lambda_2 & & & \\
& \mu_1 & & & & & \mu_3 & * & & \lambda_2 & & \\
& & \mu_1 & & & & \mu_2 & & * & \lambda_3 & \lambda_2 & \\
& & & \mu_1 & & & & \mu_2 & \mu_3 & * & & \lambda_2 \\
& & & & \mu_1 & & & & \mu_2 & & * & \lambda_3 \\
\mu & & & & & \mu_1 & & & & \mu_2 & \mu_3 & *
\end{array}
\right) ;
$$

at level 2, we have a (6×6) block matrix with blocks of order 2:

$$
Q = \left(
\begin{array}{cc|cc|cc|cc|cc|cc}
* & \lambda_3 & \lambda_2 & & & & \lambda_1 & & & & & \\
\mu_3 & * & & \lambda_2 & & & & \lambda_1 & & & & \\
\hline
\mu_2 & & * & \lambda_3 & \lambda_2 & & & & \lambda_1 & & & \\
& \mu_2 & \mu_3 & * & & \lambda_2 & & & & \lambda_1 & & \\
\hline
& & \mu_2 & & * & \lambda_3 & & & & & \lambda_1 & \\
& & & \mu_2 & \mu_3 & * & & & & & & \lambda_1 \\
\hline
\mu_1 & & & & & & * & \lambda_3 & \lambda_2 & & & \\
& \mu_1 & & & & & \mu_3 & * & & \lambda_2 & & \\
\hline
& & \mu_1 & & & & \mu_2 & & * & \lambda_3 & \lambda_2 & \\
& & & \mu_1 & & & & \mu_2 & \mu_3 & * & & \lambda_2 \\
\hline
& & & & \mu_1 & & & & \mu_2 & & * & \lambda_3 \\
\mu & & & & & \mu_1 & & & & \mu_2 & \mu_3 & *
\end{array}
\right) ;
$$

and finally, at level 3, we have a (12×12) block matrix with blocks of order 1 as in (1.1).

2.2 Vector–Kronecker Product Multiplication Algorithm

The algorithm we discuss next is at the heart of all iterative solvers for sums of Kronecker products. It shows how one can multiply a vector with a Kronecker product of H factors. Here we present its left-oriented version suitable for systems of equations arising from MCs, where the factors are rectangular matrices $X^{(h)} \in \mathbb{R}^{n_h \times m_h}$ for $h = 1, \ldots, H$.

The algorithm is based on the identity

$$\bigotimes_{h=1}^{H} X^{(h)} = \prod_{h=1}^{H} I_{m_1} \otimes \cdots \otimes I_{m_{h-1}} \otimes X^{(h)} \otimes I_{n_{h+1}} \otimes \cdots \otimes I_{n_H},$$

or, more simply,

$$\bigotimes_{h=1}^{H} X^{(h)} = \prod_{h=1}^{H} \left(I_{\prod_{f=1}^{h-1} m_f} \otimes X^{(h)} \otimes I_{\prod_{f=h+1}^{H} n_f} \right), \tag{2.4}$$

which is due to the *compatibility* of Kronecker product with matrix multiplication [66]. The left multiplication of $\mathbf{x} \in \mathbb{R}^{1 \times \prod_{h=1}^{H} n_h}$ with $\bigotimes_{h=1}^{H} X^{(h)}$ yields a product vector whose length ranges from $m_1 \prod_{h=2}^{H} n_h$ to $\prod_{h=1}^{H} m_h$ during the course of the multiplication. In (2.4), the hth factor of the form $I_{\prod_{f=1}^{h-1} m_f} \otimes X^{(h)} \otimes I_{\prod_{f=h+1}^{H} n_f}$ is a rectangular ($\prod_{f=1}^{h-1} m_f \prod_{f=h}^{H} n_f \times \prod_{f=1}^{h} m_f \prod_{f=h+1}^{H} n_f$) block diagonal matrix having $\prod_{f=1}^{h-1} m_f$ diagonal blocks each of size ($n_h \prod_{f=h+1}^{H} n_f \times m_h \prod_{f=h+1}^{H} n_f$). Furthermore, each of the blocks along the diagonal is an ($n_h \times m_h$) block matrix, where each subblock is a diagonal matrix of order $\prod_{f=h+1}^{H} n_f$ with a particular entry of $X^{(h)}$ appearing along its diagonal $\prod_{f=h+1}^{H} n_f$ many times. It is this feature that is used in devising the algorithm (cf. [66]).

Algorithm 1. *Vector–Kronecker product multiplication algorithm for* $\mathbf{x}' = \mathbf{x} \bigotimes_{h=1}^{H} X^{(h)}$

Copy $\mathbf{x}$ to $\mathbf{x}'$; $i_{\text{left}} = 1$; $i_{\text{right}} = \prod_{h=2}^{H} n_h$; $n_{H+1} = 1$;
For $h = 1$ to H,
    $base_i = 0$; $base_j = 0$;
    For $i_l = 0, \ldots, i_{\text{left}} - 1$,
        For $i_r = 0, \ldots, i_{\text{right}} - 1$,
            $index_i = base_i + i_r$;
            For $row = 0, \ldots, n_h - 1$,
                $\mathbf{z}(row) = \mathbf{x}'(index_i)$; $index_i = index_i + i_{\text{right}}$;
            $\mathbf{z}' = \mathbf{z}X^{(h)}$;
            $index_j = base_j + i_r$;
            For $col = 0, \ldots, m_h - 1$,
                $\mathbf{x}''(index_j) = \mathbf{z}'(col)$; $index_j = index_j + i_{\text{right}}$;
        $base_i = base_i + n_h i_{\text{right}}$; $base_j = base_j + m_h i_{\text{right}}$;
    $i_{\text{left}} = i_{\text{left}} m_h$; $i_{\text{right}} = i_{\text{right}} / n_{h+1}$;
    Copy $\mathbf{x}''$ to $\mathbf{x}'$.

The only floating-point arithmetic operations (flops) in Algorithm 1 take place in the vector–matrix multiplication $\mathbf{z}' = \mathbf{z}X^{(h)}$, which can be simplified when $X^{(h)} = I_{n_h \times m_h}$. The remaining operations are index manipulation and copying of vector entries. With regard to space, the algorithm requires two temporary floating-point vectors, $\mathbf{z}$ and $\mathbf{z}'$, that need to be as long as $\max_h(n_h)$ and $\max_h(m_h)$, respectively, and two floating-point vectors, $\mathbf{x}'$ and $\mathbf{x}''$, of length $\max_{h=0,\ldots,H}(\prod_{f=1}^h m_f \prod_{f=h+1}^H n_f)$ to compute and return the result. When $X^{(h)} = I_{n_h \times m_h}$, entries of $\mathbf{x}'$ can be directly copied to $\mathbf{x}''$ using appropriate index manipulations. Furthermore, in programming languages that provide handles to memory addresses in the form of pointers, it is possible to carry out the last statement by swapping the two pointers to the arrays $\mathbf{x}'$ and $\mathbf{x}''$ since $\mathbf{x}''$ is to be rewritten in the next turn of the outer loop anyway.

The complexity of a vector multiplication with $\bigotimes_{h=1}^H X^{(h)}$ amounts to

$$2 \sum_{h=1}^H nz_{X^{(h)}} \prod_{f=1}^{h-1} m_f \prod_{f=h+1}^H n_f$$

flops, where $nz_{X^{(h)}}$ is the number of nonzeros in $X^{(h)}$ for $h = 1, \ldots, H$. Therefore, the complexity of a vector multiplication with $\sum_{k=1}^K \bigotimes_{h=1}^H X_k^{(h)}$, where $X_k^{(h)} \in \mathbb{R}^{n_h \times m_h}$ for $k = 1, \ldots, K$ and $h = 1, \ldots, H$, becomes

$$K \prod_{h=1}^H m_h + 2 \sum_{k=1}^K \sum_{h=1}^H nz_{X_k^{(h)}} \prod_{f=1}^{h-1} m_f \prod_{f=h+1}^H n_f.$$

Here, the first term is due to the summation of the K product vectors obtained through Algorithm 1. Note that this expression can be simplified further for Q_O in which the factors are square by substituting n_h for m_h. In that case, one obtains $n(K + 2\sum_{h=1}^H nz_{Q^{(h)}}/n_h)$, where $nz_{Q^{(h)}} = \sum_k nz_{Q_k^{(h)}}$ and $nz_{Q_k^{(h)}}$ is the number of nonzeros in $Q_k^{(h)}$ for $k = 1, \ldots, K$ and $h = 1, \ldots, H$.

As was mentioned previously, functional transitions are a feature of SANs and are said to generalize a Kronecker product. Functional transitions enable the elegant modeling of dependencies among subsystems; yet they are not easy to implement efficiently. There is a vector-generalized Kronecker product multiplication algorithm [66] that may be used in the presence of functional transitions. For instance, in the PEPS tool where it is implemented, the different values the functions can take are hard coded into the package. This is done for each problem under consideration. Whenever a function needs to be evaluated in a particular state during the vector-generalized Kronecker product multiplication, the code jumps to a specific location dictated by the problem and carries out a sequence of if statements, trying to find out what the function evaluates to for that state. In practice, this process slows down the code considerably but does not come across when one looks at the time complexity result of vector-generalized Kronecker product multiplication.

In the next section, we discuss preprocessing techniques to expedite the analysis of MCs based on Kronecker products.

2.3 Preprocessing

Three techniques can be used to put the Kronecker representation into a more favorable form before solvers take over. These are reordering, grouping, and lumping. We discuss them in order.

We assume that Q represents a time-homogeneous CTMC. Therefore, the left-hand side of each equation in (2.1) is constant up to a symmetric permutation, that is, up to a reordering of the state space, S. Yet, there may be multiple ways in which the number of Kronecker product terms, K, and the number of factors in each Kronecker product, H, forming Q_O in (2.1) are chosen. Obviously, the choice $(K, H) = (1, 1)$ indicates a flat representation and is assumed to be impossible due to memory limitations. As H decreases toward one, the Kronecker representation becomes flatter, implying increased storage requirements. On the other hand, if K were one, then Q could be analyzed along each dimension independently. Hence, K is normally assumed to be larger than one. Observe that it would be advantageous to be able to make K as small as possible without changing H since then we would be decreasing the number of terms in the Kronecker representation of Q_O and making the $Q_k^{(h)}$ matrices denser.

Reordering in MCs based on Kronecker products refers to either permuting the factors of Kronecker products or renumbering the states in the state spaces of factors. The latter corresponds to a symmetric permutation of the factor matrices $Q_k^{(h)}$ for $k = 1, \ldots, K$ associated with the renumbered state space $S^{(h)}$. As indicated in [10,67], reordering of the first kind may be used to reduce the overhead associated with vector–Kronecker product multiplication in the presence of functional transitions. Furthermore, reordering of both kinds can change the nonzero structure of the underlying MC and thereby have an effect on the convergence of iterative methods sensitive to the nonzero structure [52]. Hence, with the help of reordering, it may be possible to symmetrically permute the nonzero structure of the underlying MC to a more favorable form for the iterative method of choice. In doing this, we can use the nonzero structure of $\sum_{k=1}^{K} Q_k^{(h)}$, which indicates how factor h contributes to the nonzero structure of Q_O for $h = 1, \ldots, H$.

Grouping in MCs based on Kronecker products refers to collapsing the same adjacent factors in each Kronecker product. Consequently, the factors in each Kronecker product are reduced by the same number and the state-space sizes of the factors are increased. The effect of grouping factors in Kronecker products forming Q_O is investigated in a sequence of papers [10,66,67] on functional transitions. The objective is to reduce the number of factors and, thereby, the overhead associated with evaluating functional transitions. Results show that in some cases grouping may help to reduce the state-space size if it had unreachable states, may decrease the overhead associated with functional transitions, and may even decrease the number of terms in the Kronecker representation. When there are functional transitions, the best approach seems to be to group those factors that have functional dependencies among each other. In the absence of functional transitions, it is recommended to group as many factors as possible given available memory starting from the highest indexed factor. This ensures a flatter representation for diagonal blocks at a particular level, which is a useful feature in certain iterative methods.

The effects of reordering and grouping of factors of Kronecker products on the convergence and space requirements of iterative methods have been investigated in a number of papers [32, 34, 54, 85, 146], but a broad, systematic study seems to be lacking.

Lumpability [93] is a property possessed by some MCs that, if conditions are met, may be used to reduce a large state space to a smaller one. The idea is to find a partitioning of the original state space such that, when the states in each partition are lumped (or *aggregated*) to form a single state, the resulting MC described by the lumped states has equivalent behavior to the original chain. It is therefore important to consider lumping to reduce the size of the state space, S, before moving to the solution phase.

In this work we refer to two kinds of lumpability: ordinary lumpability and exact lumpability. Here we give definitions for CTMCs. Equivalent definitions can be stated for DTMCs. A CTMC represented by Q is said to be *ordinarily lumpable* with respect to a partitioning of its state space $S = \bigcup_l S_l$ and $S_l \cap S_u = \emptyset$ for all $l \neq u$ if for all $S_l \subset S$ and all $\mathbf{i}, \mathbf{i}' \in S_l$

$$\sum_{\mathbf{j} \in S_u} q(\mathbf{i}, \mathbf{j}) = \sum_{\mathbf{j} \in S_u} q(\mathbf{i}', \mathbf{j}) \text{ for all } S_u \subset S. \tag{2.5}$$

A CTMC represented by Q is said to be *exactly lumpable* with respect to a partitioning of its state space $S = \bigcup_l S_l$ and $S_l \cap S_u = \emptyset$ for all $l \neq u$ if for all $S_l \subset S$ and all $\mathbf{i}, \mathbf{i}' \in S_l$

$$\sum_{\mathbf{j} \in S_u} q(\mathbf{j}, \mathbf{i}) = \sum_{\mathbf{j} \in S_u} q(\mathbf{j}, \mathbf{i}') \text{ for all } S_u \subset S. \tag{2.6}$$

Ordinary lumpability refers to a partitioning of the state space in which the sums of transition rates from each state in a partition to a(nother) partition are the same. On the other hand, exact lumpability refers to a partitioning of the state space in which the sums of transition rates from all states in a partition into each state of a(nother) partition are the same.

Let S_{lumped} denote the lumped state space. On an ordinarily lumped MC one can only compute performance measures defined over S_{lumped}. On an exactly lumped MC one can compute steady-state performance measures defined over S, transient performance measures defined over S_{lumped}, and transient performance measures defined over S if the states in the exactly lumpable partitions have the same initial probabilities. Since MCs satisfy a row sum property rather than a column sum property, the exact lumpability condition in (2.6) is harder to satisfy than the ordinary lumpability condition in (2.5). See [20] and references therein for more information regarding the concept of lumpability and its implications.

Lumpability can be investigated on the flat representation of a MC. Detection of ordinary and exact lumpability on Q through partition refinement would imply a time complexity of $O(nz \log n)$ and a space complexity of $O(nz)$ [116]. Since this is expensive in terms of time and storage, techniques that investigate lumpability on the Kronecker representation have been considered.

Lumpability can be investigated within each of the state spaces $\mathcal{S}^{(h)}$ that define the Kronecker representation of Q_O in (2.1) for $h = 1, \ldots, H$ independently. For the state space $\mathcal{S}^{(h)}$, detection of ordinary and exact lumpability through partition refinement as in [29] requires a time complexity of $O(nz_{Q^{(h)}} \log n_h)$ and a space complexity of $O(nz_{Q^{(h)}})$. Then the lumped Kronecker representation may be obtained by replacing each of the state spaces $\mathcal{S}^{(h)}$ and its corresponding matrices $Q_k^{(h)}$ for $k = 1, \ldots, K$ with equivalent lumped ones. Lumpability can also be investigated among the state spaces $\mathcal{S}^{(h)}$ that are *replicated* (or identical) with respect to the Kronecker representation of Q_O as in [9]. There, ordinary lumpability of replicated state spaces is shown in the presence of functional transitions. Note that replication refers to a very specific kind of symmetry in the Kronecker representation, and with ordinary lumpability only performance measures of interest over $\mathcal{S}_{\text{lumped}}$ can be computed. Lumpability can also be investigated among the state spaces $\mathcal{S}^{(h)}$ by considering dependencies and matrix properties in the Kronecker representation as in [87, 88]. There, sufficient conditions that satisfy ordinary lumpability are specified and an iterative steady-state solution method that is able to compute performance measures over $\mathcal{S}$ is given for CTMCs and DTMCs in the presence of functional transitions. The work identifies lumpable partitionings on the underlying MC induced by the nested block structure of the Kronecker representation in (2.2). Although the particular approach of lumping one or more state spaces $\mathcal{S}^{(h)}$ totally as in [87, 88] is a very specific kind of performance equivalence and lumping considered in [21, 23], due to its accommodation of functional transitions, it also enables the detection of certain ordinarily lumpable partitionings in which blocks are composed of multiple (nonidentical) state spaces, but the individual state spaces cannot be lumped by themselves. This is not possible with the approaches in [9, 21, 23].

We remark that neither of the two approaches in [9] and [87, 88] that investigate lumpability among the state spaces $\mathcal{S}^{(h)}$ for $h = 1, \ldots, H$ is completely automated, uses a Kronecker representation for the lumped MC, and possesses a proper complexity analysis. Furthermore, since the Kronecker representation is rich in structure and the three approaches presented in this chapter do not work on the flat representation, there can very well be other symmetries in the Kronecker representation that also lead to lumpability. Along a slightly different line, the concept of *quasilumpability* [58] can be investigated.

Chapter 3
Iterative Methods

Classical iterative methods for the solution of a linear system of equations as in (1.3) start with an initial approximation. At each iteration, they multiply the current approximation with a particular matrix to obtain a new approximation with the objective that the approximations eventually converge to the true solution [83, 130]. These methods are the building blocks of all advanced iterative methods. The matrix used in the iterative multiplication process is obtained at the outset by splitting the coefficient matrix of the linear system, which is Q in our setting. Therefore, we begin by splitting the smaller matrices that form the Kronecker products as in [146] and show how classical iterative methods can be formulated in terms of these smaller matrices. We present block versions of the methods since point versions follow from the block versions by considering blocks of order one.

We continue the discussion with *projection* (or *Krylov subspace*) methods for MCs based on Kronecker products in which approximate solutions satisfying various constraints are extracted from low-dimensional subspaces [5, 128, 139]. Being iterative, their basic operation is also vector–Kronecker product multiplication. However, compared to block iterative methods, they require a larger number of supplementary vectors as long as the state-space size. But more importantly, they need to be used with preconditioners to result in effective solvers. Fortunately, the first term of the splitting associated with block iterative methods can be used as preconditioner [34].

In [19, 24–26], aggregation–disaggregation steps are coupled with various iterative methods for MCs based on Kronecker products to accelerate convergence. An *iterative aggregation–disaggregation* (IAD) method for MCs based on Kronecker products and its adaptive version, which analyzes aggregated systems for those parts where the error is estimated to be high, are proposed in [22] and [27], respectively. The adaptive IAD method in [27] is improved upon in [28] through a recursive definition and called *multilevel* (ML). Here, we present this simple ML method and then discuss a class of ML methods based on it that are shown to be quite

T. Dayar, *Analyzing Markov Chains using Kronecker Products: Theory and Applications*, SpringerBriefs in Mathematics, DOI 10.1007/978-1-4614-4190-8_3,
© Tuğrul Dayar 2012

effective [33, 35] in solving a large number of problems in the literature. ML methods can easily use iterative methods based on splittings at each level before aggregation and after disaggregation.

3.1 Splitting the Smaller Matrices

Let

$$Q_k^{(h)} = D_k^{(h)} + U_k^{(h)} + L_k^{(h)} \quad \text{for} \quad k = 1, \ldots, K \quad \text{and} \quad h = 1, \ldots, H, \quad (3.1)$$

where $D_k^{(h)}$, $U_k^{(h)}$, and $L_k^{(h)}$ are respectively the diagonal, strictly upper-triangular, and strictly lower-triangular parts of $Q_k^{(h)}$. Observe that $D_k^{(h)} \geq 0$, $U_k^{(h)} \geq 0$, and $L_k^{(h)} \geq 0$ since $Q_k^{(h)} \geq 0$. Then using Lemma A.8 in [146], which rests on the associativity of Kronecker product and the *distributivity* of Kronecker product over matrix addition, it is possible to express Q_O of Q in (2.1) at level $l = 0, \ldots, H$ using (3.1) as

$$Q_O = Q_{U(l)} + Q_{L(l)} + Q_{DU(l)} + Q_{DL(l)}, \quad (3.2)$$

where

$$Q_{U(l)} = \sum_{k=1}^{K} \sum_{h=1}^{l} \left(\bigotimes_{f=1}^{h-1} D_k^{(f)} \right) \otimes U_k^{(h)} \otimes \left(\bigotimes_{f=h+1}^{H} Q_k^{(f)} \right), \quad (3.3)$$

$$Q_{L(l)} = \sum_{k=1}^{K} \sum_{h=1}^{l} \left(\bigotimes_{f=1}^{h-1} D_k^{(f)} \right) \otimes L_k^{(h)} \otimes \left(\bigotimes_{f=h+1}^{H} Q_k^{(f)} \right) \quad (3.4)$$

correspond respectively to the strictly block upper- and lower-triangular parts of Q_O at level l, and

$$Q_{DU(l)} = \sum_{k=1}^{K} \sum_{h=l+1}^{H} \left(\bigotimes_{f=1}^{h-1} D_k^{(f)} \right) \otimes U_k^{(h)} \otimes \left(\bigotimes_{f=h+1}^{H} Q_k^{(f)} \right), \quad (3.5)$$

$$Q_{DL(l)} = \sum_{k=1}^{K} \sum_{h=l+1}^{H} \left(\bigotimes_{f=1}^{h-1} D_k^{(f)} \right) \otimes L_k^{(h)} \otimes \left(\bigotimes_{f=h+1}^{H} Q_k^{(f)} \right) \quad (3.6)$$

correspond respectively to the strictly upper- and lower-triangular parts of the block diagonal of Q_O at level l. Observe that $Q_{U(l)} \geq 0$, $Q_{L(l)} \geq 0$, $Q_{DU(l)} \geq 0$, and $Q_{DL(l)} \geq 0$. Furthermore, $l = 0$ implies Q_O is a single block for which $Q_{U(0)} = Q_{L(0)} = 0$, whereas $l = H$ corresponds to a pointwise partitioning of Q_O for which $Q_{DU(H)} = Q_{DL(H)} = 0$. Hence, for iterative methods based on *block partitionings*, $l = 1, \ldots, H - 1$ should be used.

Example 1 (ctnd.). Consider the block partitioning of our problem at level 0 for which $l = 0$, $b_0 = 1$, and Q_O is viewed as a (1×1) block matrix with blocks of order $o_0 = 12$ [see (2.2)]. Then, from (3.3) and (3.4) we have

$$Q_{U(0)} = 0, \quad Q_{L(0)} = 0, \quad Q_{U(0)} + Q_{L(0)} = 0,$$

whereas from (3.5) and (3.6) we have

$$Q_{DU(0)} = \sum_{k=1}^{4} U_k^{(1)} \otimes Q_k^{(2)} \otimes Q_k^{(3)} + \sum_{k=1}^{4} D_k^{(1)} \otimes U_k^{(2)} \otimes Q_k^{(3)}$$

$$+ \sum_{k=1}^{4} D_k^{(1)} \otimes D_k^{(2)} \otimes U_k^{(3)},$$

$$Q_{DL(0)} = \sum_{k=1}^{4} L_k^{(1)} \otimes Q_k^{(2)} \otimes Q_k^{(3)} + \sum_{k=1}^{4} D_k^{(1)} \otimes L_k^{(2)} \otimes Q_k^{(3)}$$

$$+ \sum_{k=1}^{4} D_k^{(1)} \otimes D_k^{(2)} \otimes L_k^{(3)},$$

$$Q_{DU(0)} + Q_{DL(0)} = Q_O.$$

Note that there is $\sqrt{b_0} = 1$ block along the diagonal.

Now consider the block partitioning of the problem at level 1 for which $l = 1$, $b_1 = 4$, and Q_O is viewed as a (2×2) block matrix with blocks of order $o_1 = 6$ [see (2.2)]. Then, from (3.3) and (3.4) we have

$$Q_{U(1)} = \sum_{k=1}^{4} U_k^{(1)} \otimes Q_k^{(2)} \otimes Q_k^{(3)}, \quad Q_{L(1)} = \sum_{k=1}^{4} L_k^{(1)} \otimes Q_k^{(2)} \otimes Q_k^{(3)},$$

$$Q_{U(1)} + Q_{L(1)} = \begin{pmatrix} & & & & \lambda_1 & & & & & \\ & & & & & \lambda_1 & & & & \\ & & & & & & \lambda_1 & & & \\ & & & & & & & \lambda_1 & & \\ & & & & & & & & \lambda_1 & \\ & & & & & & & & & \lambda_1 \\ \hline \mu_1 & & & & & & & & & \\ & \mu_1 & & & & & & & & \\ & & \mu_1 & & & & & & & \\ & & & \mu_1 & & & & & & \\ & & & & \mu_1 & & & & & \\ \mu & & & & & \mu_1 & & & & \end{pmatrix},$$

whereas from (3.5) and (3.6) we have

$$Q_{DU(1)} = \sum_{k=1}^{4} D_k^{(1)} \otimes U_k^{(2)} \otimes Q_k^{(3)} + \sum_{k=1}^{4} D_k^{(1)} \otimes D_k^{(2)} \otimes U_k^{(3)},$$

$$Q_{DL(1)} = \sum_{k=1}^{4} D_k^{(1)} \otimes L_k^{(2)} \otimes Q_k^{(3)} + \sum_{k=1}^{4} D_k^{(1)} \otimes D_k^{(2)} \otimes L_k^{(3)},$$

$$Q_{DU(1)} + Q_{DL(1)} = \left(\begin{array}{ccccc|ccccc}
\lambda_3 & \lambda_2 & & & & & & & & \\
\mu_3 & & \lambda_2 & & & & & & & \\
\mu_2 & & \lambda_3 & \lambda_2 & & & & & & \\
& \mu_2 & \mu_3 & & \lambda_2 & & & & & \\
& & \mu_2 & & \lambda_3 & & & & & \\
& & & \mu_2 & \mu_3 & & & & & \\
\hline
& & & & & \lambda_3 & \lambda_2 & & & \\
& & & & & \mu_3 & & \lambda_2 & & \\
& & & & & \mu_2 & & \lambda_3 & \lambda_2 & \\
& & & & & & \mu_2 & \mu_3 & & \lambda_2 \\
& & & & & & & \mu_2 & & \lambda_3 \\
& & & & & & & & \mu_2 & \mu_3 \\
\end{array} \right).$$

Note that there are $\sqrt{b_1} = 2$ blocks along the diagonal.

Finally, consider the block partitioning of the same problem at level 2 for which $l = 2$, $b_2 = 36$, and Q_O is viewed as a (6×6) block matrix with blocks of order $o_2 = 2$ [see (2.2)]. Then, from (3.3) and (3.4) we have

$$Q_{U(2)} = \sum_{k=1}^{4} U_k^{(1)} \otimes Q_k^{(2)} \otimes Q_k^{(3)} + \sum_{k=1}^{4} D_k^{(1)} \otimes U_k^{(2)} \otimes Q_k^{(3)},$$

$$Q_{L(2)} = \sum_{k=1}^{4} L_k^{(1)} \otimes Q_k^{(2)} \otimes Q_k^{(3)} + \sum_{k=1}^{4} D_k^{(1)} \otimes L_k^{(2)} \otimes Q_k^{(3)},$$

$$Q_{U(2)} + Q_{L(2)} = \left(\begin{array}{cc|cc|cc|cc|cc|cc}
& \lambda_2 & & & \lambda_1 & & & & & & & \\
& & \lambda_2 & & & \lambda_1 & & & & & & \\
\mu_2 & & & & \lambda_2 & & & \lambda_1 & & & & \\
& \mu_2 & & & & \lambda_2 & & & \lambda_1 & & & \\
& & \mu_2 & & & & & & & \lambda_1 & & \\
& & & \mu_2 & & & & & & & \lambda_1 & \\
\hline
\mu_1 & & & & & & & \lambda_2 & & & & \\
& \mu_1 & & & & & & & \lambda_2 & & & \\
& & \mu_1 & & & & \mu_2 & & & \lambda_2 & & \\
& & & \mu_1 & & & & \mu_2 & & & \lambda_2 & \\
& & & & \mu_1 & & & & \mu_2 & & & \\
\mu & & & & & \mu_1 & & & & \mu_2 & & \\
\end{array} \right),$$

whereas from (3.5) and (3.6) we have

$$Q_{DU(2)} = \sum_{k=1}^{4} D_k^{(1)} \otimes D_k^{(2)} \otimes U_k^{(3)},$$

$$Q_{DL(2)} = \sum_{k=1}^{4} D_k^{(1)} \otimes D_k^{(2)} \otimes L_k^{(3)},$$

$$Q_{DU(2)} + Q_{DL(2)} = \begin{pmatrix} \lambda_3 & & & & & & & \\ \mu_3 & & & & & & & \\ & \lambda_3 & & & & & & \\ & \mu_3 & & & & & & \\ & & \lambda_3 & & & & & \\ & & \mu_3 & & & & & \\ & & & \lambda_3 & & & & \\ & & & \mu_3 & & & & \\ & & & & \lambda_3 & & & \\ & & & & \mu_3 & & & \\ & & & & & \lambda_3 & \\ & & & & & \mu_3 \end{pmatrix}.$$

There are $\sqrt{b_2} = 6$ blocks along the diagonal.

At $l = 3$, Q_O is a (12×12) block matrix with blocks of order one.

3.2 Block Iterative Methods

Now let Q in (2.1) be irreducible and *split* at level l using (3.2) as

$$Q = Q_O + Q_D = Q_{U(l)} + Q_{L(l)} + Q_{DU(l)} + Q_{DL(l)} + Q_D = M - N, \quad (3.7)$$

where M is nonsingular (i.e., M^{-1} exists). Then, *power, block Jacobi overrelaxation* (BJOR) and *block successive overrelaxation* (BSOR) methods are based on different splittings of Q, and each satisfies

$$\pi_{(m+1)} M = \pi_{(m)} N \quad \text{for} \quad m = 0, 1, \ldots$$

with the sequence of approximations $\pi_{(m+1)}$ to the steady-state vector in (1.3), where $\pi_{(0)} > 0$ is the initial approximation such that $\pi_{(0)} e = 1$ and $T = NM^{-1}$ is the *iteration matrix*. Note that T does not change from iteration to iteration, and only the current approximation is used to compute the new approximation. Hence, these methods based on splittings of the coefficient matrix are also known as *stationary* iterative methods. Since Q is a singular matrix and assumed to be irreducible, the largest eigenvalue [78, 111] of T in magnitude is one. To ensure convergence,

T should not have other eigenvalues with magnitude one. For converging approximations, the magnitude of the eigenvalue of T closest to one determines the *rate of convergence* [5, 139].

The particular splittings corresponding to power, BJOR, and *(forward)* BSOR methods are

$$M^{\text{Power}} = -\alpha I,$$

$$N^{\text{Power}} = -\alpha \left(I + \frac{1}{\alpha} Q \right),$$

$$M^{\text{BJOR}} = \frac{1}{\omega}(Q_D + Q_{DU(l)} + Q_{DL(l)}),$$

$$N^{\text{BJOR}} = \frac{1-\omega}{\omega}(Q_D + Q_{DU(l)} + Q_{DL(l)}) - Q_{U(l)} - Q_{L(l)},$$

$$M^{\text{BSOR}} = \frac{1}{\omega}(Q_D + Q_{DU(l)} + Q_{DL(l)}) + Q_{U(l)},$$

$$N^{\text{BSOR}} = \frac{1-\omega}{\omega}(Q_D + Q_{DU(l)} + Q_{DL(l)}) - Q_{L(l)},$$

where $\alpha \in [\max_{s \in \mathcal{S}} |q_D(s, s)|, \infty)$ is the *uniformization parameter* of the power method and $\omega \in (0, 2)$ is the *relaxation parameter* of the BJOR and BSOR methods. Here, forward iteration refers to computing unknowns ordered toward the beginning of the state space earlier than unknowns ordered later in the state space. The power method works at level $l = H$ since it is a point method. Furthermore, BJOR and BSOR reduce to the *block Jacobi* (BJacobi) and *block Gauss–Seidel* (BGS) methods for $\omega = 1$, and they become point JOR and point SOR methods for $l = H$. Note that [86] shows how one can find $\max_{s \in \mathcal{S}} |q_D(s, s)|$ in the presence of functional transitions when Q_D is given as a sum of Kronecker products.

Since $Q = Q_O + Q_D$, the power method at iteration m can be expressed as

$$\pi_{(m+1)} = \pi_{(m)} + \frac{1}{\alpha} \pi_{(m)} Q_D + \frac{1}{\alpha} \pi_{(m)} Q_O. \tag{3.8}$$

The second term in (3.8) poses no problem from a computational point of view since Q_D is diagonal, and the third term can be efficiently implemented using the vector–Kronecker product multiplication algorithm since Q_O is a sum of Kronecker products [see (2.1)].

The BJOR method with a level l block partitioning at iteration m is

$$\pi_{(m+1)}(Q_D + Q_{DU(l)} + Q_{DL(l)}) = (1-\omega)\pi_{(m)}Q_D + (1-\omega)\pi_{(m)}Q_{DU(l)}$$
$$+ (1-\omega)\pi_{(m)}Q_{DL(l)}$$
$$- \omega\pi_{(m)}Q_{U(l)} - \omega\pi_{(m)}Q_{L(l)}.$$
$$\tag{3.9}$$

This is a block diagonal linear system with $\sqrt{b_l}$ blocks of order o_l along the diagonal of the nonsingular coefficient matrix $(Q_D + Q_{DU(l)} + Q_{DL(l)})$ and a nonzero right-hand side that can be efficiently computed using the vector–Kronecker product multiplication algorithm since $Q_{U(l)}$, $Q_{L(l)}$, $Q_{DU(l)}$, and $Q_{DL(l)}$ are sums of Kronecker products [see (3.3)–(3.6)]. Hence, (3.9) is equivalent to $\sqrt{b_l}$ independent, nonsingular linear systems each of order o_l and a nonzero right-hand side. If there is space, one can generate and factorize in sparse storage nonsingular blocks of the form

$$
Q((i_1, \ldots, i_l), (i_1, \ldots, i_l)) = \sum_{k=1}^{K} \left(\prod_{h=1}^{l} q_k^{(h)}(i_h, i_h) \right) \left(\bigotimes_{h=l+1}^{H} Q_k^{(h)} \right)
$$

$$
+ \, Q_D((i_1, \ldots, i_l), (i_1, \ldots, i_l))
$$

$$
\text{for} \quad (i_1, \ldots, i_l) \in \times_{h=1}^{l} \mathcal{S}^{(h)} \tag{3.10}
$$

along the diagonal [see (2.3)] of $(Q_D + Q_{DU(l)} + Q_{DL(l)})$ at the outset and solve the $|\times_{h=1}^{l} \mathcal{S}^{(h)}| = \sqrt{b_l}$ systems directly at each iteration. Otherwise, one can use an iterative method, even a block iterative method, such as BJOR, since the off-diagonal parts of diagonal blocks given by

$$
\sum_{k=1}^{K} \left(\prod_{h=1}^{l} q_k^{(h)}(i_h, i_h) \right) \left(\bigotimes_{h=l+1}^{H} Q_k^{(h)} \right)
$$

are sums of Kronecker products.

The situation with the BSOR method is not very different from that of BJOR. For BSOR with a level l block partitioning, at iteration m we have

$$
\pi_{(m+1)}(Q_D + Q_{DU(l)} + Q_{DL(l)} + \omega Q_{U(l)})
$$

$$
= (1 - \omega)\pi_{(m)} Q_D + (1 - \omega)\pi_{(m)} Q_{DU(l)}
$$

$$
+ (1 - \omega)\pi_{(m)} Q_{DL(l)} - \omega \pi_{(m)} Q_{L(l)}. \tag{3.11}
$$

This is a block upper-triangular linear system with $\sqrt{b_l}$ blocks of order o_l along the diagonal of the nonsingular coefficient matrix $(Q_D + Q_{DU(l)} + Q_{DL(l)} + \omega Q_{U(l)})$ and a nonzero right-hand side that can be efficiently computed using the vector–Kronecker product multiplication algorithm since $Q_{L(l)}$, $Q_{DU(l)}$, and $Q_{DL(l)}$ are sums of Kronecker products. In [146], a recursive algorithm is given for a nonsingular linear system with a lower-triangular coefficient matrix in the form of a sum of Kronecker products and a nonzero right-hand side. Such a system arises in *backward* point SOR. There, a version of the same algorithm for backward BSOR is also discussed. Here we remark that a nonrecursive block upper-triangular solution algorithm for (3.11) is also possible [32] and a block row-oriented version is preferable in the presence of functional transitions:

Algorithm 2. *Nonrecursive block upper-triangular solution at level l for MCs based on Kronecker products*

$\mathbf{b} = (1 - \omega)\boldsymbol{\pi}_{(m)} Q_D + (1 - \omega)\boldsymbol{\pi}_{(m)} Q_{DU(l)} + (1 - \omega)\boldsymbol{\pi}_{(m)} Q_{DL(l)} - \omega\boldsymbol{\pi}_{(m)} Q_{L(l)};$
For row of blocks $(i_1, \ldots, i_l) = (0, \ldots, 0)$ to $(n_1 - 1, \ldots, n_l - 1)$
lexicographically,
    Solve $\boldsymbol{\pi}_{(m+1)}((i_1, \ldots, i_l)) Q((i_1, \ldots, i_l), (i_1, \ldots, i_l)) = \mathbf{b}((i_1, \ldots, i_l));$
    For column of blocks $(j_1, \ldots, j_l) > (i_1, \ldots, i_l),$
       $\mathbf{b}((j_1, \ldots, j_l)) = \mathbf{b}((j_1, \ldots, j_l))$
          $-\omega\boldsymbol{\pi}_{(m+1)}((i_1, \ldots, i_l)) Q_{U(l)}((i_1, \ldots, i_l), (j_1, \ldots, j_l)).$

In Algorithm 2, initially the nonzero right-hand-side vector $\mathbf{b}$ can be efficiently computed using the vector–Kronecker product multiplication algorithm since $Q_{L(l)}$, $Q_{DU(l)}$, and $Q_{DL(l)}$ are sums of Kronecker products. Furthermore, $Q((i_1, \ldots, i_l), (i_1, \ldots, i_l))$ is given in (3.10) in terms of a sum of Kronecker products, and $Q_{U(l)}((i_1, \ldots, i_l), (j_1, \ldots, j_l))$ for $(j_1, \ldots, j_l) > (i_1, \ldots, i_l)$ can be expressed in terms of a sum of Kronecker products using (3.3) as

$$Q((i_1, \ldots, i_l), (j_1, \ldots, j_l))$$

$$= \sum_{k=1}^{K} \sum_{h=1}^{l} \left(\prod_{f=1}^{h-1} d_k^{(f)}(i_f, i_f) \right) u_k^{(h)}(i_h, j_h) \left(\prod_{f=h+1}^{l} q_k^{(f)}(i_f, j_f) \right) \left(\bigotimes_{f=l+1}^{H} Q_k^{(f)} \right).$$

In contrast to BJOR, the nonsingular diagonal blocks $Q((i_1, \ldots, i_l), (i_1, \ldots, i_l))$ in BSOR must be solved in lexicographical order. If there is space, one can generate and factorize in sparse storage these blocks as in BJOR at the outset and solve the $\sqrt{b_l}$ systems directly at each iteration. Otherwise, one can use an iterative method such as BSOR since the off-diagonal parts of diagonal blocks are also sums of Kronecker products. After each block is solved for the unknown subvector $\boldsymbol{\pi}_{(m+1)}((i_1, \ldots, i_l))$, $\mathbf{b}$ is updated by multiplying the computed subvector by the corresponding row of blocks above the diagonal. Finally, we emphasize that BSOR at level l reduces to point SOR if $Q_{DL(l)} = 0$ (Remark 4.1 in [146]).

Surprisingly, block iterative solvers, which are sometimes called two-level (or two-stage) iterative solvers [112], have still not been incorporated into most analysis packages based on Kronecker representations, although they have been shown to be more effective than point solvers on many test cases [32, 146]. Furthermore, in contrast to the block partitionings of sparse MCs considered in [59], block partitionings of Kronecker products are nested and recursive due to the lexicographical ordering of states. Therefore, there tends to be more common structure among the diagonal blocks of a MC expressed as a sum of Kronecker products. Diagonal blocks having identical off-diagonal parts and diagonals that differ by a multiple of the identity are exploited in [32]. There, it is shown that such diagonal blocks can share and work with the factorization of only one diagonal block. This approach saves not only time spent for factorization of diagonal blocks at the outset but also space. The same paper also discusses a three-level version of BSOR for MCs based on Kronecker products in which diagonal blocks that are too large to be factorized are solved using BSOR. Similar results also appear in [85]. Finally, it is possible to alter the nonzero structure of the underlying MC of a Kronecker representation by reordering factors and states of factors so as to make it more suitable for block iterative methods. Obviously, power and point JOR methods will not benefit from such reordering.

The next section discusses various preconditioners to be used with projection methods for MCs based on Kronecker products.

3.3 Preconditioned Projection Methods

Projection methods for MCs [5, 59, 128, 139] are *nonstationary* iterative methods [83, 130] using a larger number of supplementary vectors than block iterative methods to expedite the solution process. The most commonly used projection methods for the solution of nonsymmetric linear systems are *biconjugate gradient* (BCG) [68], *generalized minimal residual* (GMRES) [131], *conjugate gradient squared* (CGS) [137], *quasiminimal residual* (QMR) [76], and *biconjugate gradient stabilized* (BICGSTAB) [147]. Of these, GMRES uses as many supplementary vectors as the Krylov subspace [83, 130] size and therefore has the highest memory requirements.

Projection methods need to be used with preconditioners [12] to result in effective solvers. At each iteration of a preconditioned projection method, the row residual vector, $\mathbf{r}$, is used as the right-hand side of the linear system

$$\mathbf{z}M = \mathbf{r} \qquad (3.12)$$

to compute the preconditioned row residual vector, $\mathbf{z}$. The objective of this preconditioning step is to improve the error in the approximate solution vector at that iteration. If M were a multiple of I [as in (3.8)], then the preconditioned residual would be a multiple of the residual computed at that iteration, implying no improvement. Hence, the preconditioner should approximate the coefficient matrix of the original system in a better way, yet the solution of linear systems as in (3.12) involving the *preconditioner* matrix, M, should be cheap. It is shown in [59] through a large number of numerical experiments on benchmark problems that, to yield effective solvers, projection methods for sparse MCs should be used with preconditioners, such as those based on incomplete LU (ILU) factorizations [59, 129]. However, it is still not clear how one could devise ILU-type preconditioners for infinitesimal generators that are in the form of (2.1).

So far, various preconditioners have been proposed for Kronecker structured representations such as those based on truncated Neumann series [139, 141], the cheap and separable preconditioner [25], circulant preconditioners for a specific class of problem [41], and the Kronecker sum preconditioner [144], which has been shown to work effectively on some small problems. The Kronecker product approximate preconditioner for MCs based on Kronecker products developed in a sequence of papers [98–100], although encouraging, is in the form of a prototype implementation. Numerical experiments in [25, 26, 99, 100, 141] indicate that there is still room for research regarding the development of effective preconditioners for MCs based on Kronecker products.

In introducing another class of preconditioner, we remark that each of the block iterative methods discussed in this work is actually a preconditioned power method for which the preconditioning matrix is M in (3.7). Since M is based on Kronecker products, a BSOR preconditioner exploiting this property is proposed in [34]. In contrast to the BSOR preconditioner entertained for sparse MCs in [59], the BSOR preconditioner for MCs based on Kronecker products has a rich structure induced by the lexicographical ordering of states. Through numerical experiments, it is shown in [34] that two-level BSOR-preconditioned projection methods in which the diagonal blocks are solved exactly emerge as effective solvers that are competitive with block iterative methods and ML methods.

It will be interesting to compare point JOR, BJOR, and point SOR precondition-ers as defined in (3.9) and (3.11) with existing preconditoners for MCs based on Kronecker products. Clearly, the class of ML methods proposed in [33] is another candidate for preconditioning projection methods.

In the next section, we introduce a simple version of the ML method [28, 33] for irreducible MCs based on Kronecker products that is intimately related to the well-known IAD method [40, 95, 139] but is not restricted to having two levels. A class of ML methods is then discussed in terms of the simple ML method.

3.4   Multilevel Methods

Let $\mathcal{S}_{(l)} = \times_{h=l+1}^{H} \mathcal{S}^{(h)}$ for $l = 0, \ldots, H$ and the mapping $f_{(l)} : \mathcal{S}_{(l)} \longrightarrow \mathcal{S}_{(l+1)}$ represent the *aggregation* of dimension $(l+1)$ [i.e., the state space $S^{(l+1)}$] so that the states in $\mathcal{S}_{(l)}$ are mapped to the states in $\mathcal{S}_{(l+1)}$. Note that $\mathcal{S}_{(0)} = \mathcal{S}$ and $\mathcal{S}_{(H)} = \{0\}$. Furthermore, let the aggregated CTMCs $\tilde{Q}_{(m,l)}$ with state spaces $\mathcal{S}_{(l)}$ be defined at levels $l = 1, \ldots, H$ with $\tilde{Q}_{(m,0)} = Q$ for iteration m. Finally, let the power method be used as *smoother* (or *accelerator*) before aggregation $\eta_{(m,l)}$ times and after disaggregation $\nu_{(m,l)}$ times with $\alpha_{(m,l)} \in [\max_{i_{(l)} \in \mathcal{S}_{(l)}} |\tilde{q}_{(m,l)}(i_{(l)}, i_{(l)})|, \infty)$ at level l for iteration m. Then the ML iteration matrix at level l for iteration m is given by

$$T_{(m,l)}^{ML} = \left(I + \frac{1}{\alpha_{(m,l)}} \tilde{Q}_{(m,l)} \right)^{\eta_{(m,l)}} R_{(l)} T_{(m,l+1)}^{ML} P_{x_{(m,l)}} \left(I + \frac{1}{\alpha_{(m,l)}} \tilde{Q}_{(m,l)} \right)^{\nu_{(m,l)}}$$
(3.13)

and satisfies

$$\pi_{(m+1,l)} = \pi_{(m,l)} T_{(m,l)}^{ML} \quad \text{for} \quad m = 0, 1, \ldots,$$

where

$$x_{(m,l)} = \pi_{(m,l)} \left(I + \frac{1}{\alpha_{(m,l)}} \tilde{Q}_{(m,l)} \right)^{\eta_{(m,l)}},$$
(3.14)

$$r_{(l)}(\mathbf{i}_{(l)}, \mathbf{i}_{(l+1)}) = \begin{cases} 1 & \text{if } f_{(l)}(\mathbf{i}_{(l)}) = \mathbf{i}_{(l+1)}, \\ 0 & \text{otherwise}, \end{cases} \tag{3.15}$$

$$p_{\mathbf{x}_{(m,l)}}(\mathbf{i}_{(l+1)}, \mathbf{i}_{(l)}) = \begin{cases} \dfrac{\mathbf{x}_{(m,l)}(\mathbf{i}_{(l)})}{\sum_{\mathbf{j}_{(l)} \in \mathcal{S}_{(l)} : f_{(l)}(\mathbf{j}_{(l)}) = \mathbf{i}_{(l+1)}} \mathbf{x}_{(m,l)}(\mathbf{j}_{(l)})} & \text{if } f_{(l)}(\mathbf{i}_{(l)}) = \mathbf{i}_{(l+1)}, \\ 0 & \text{otherwise}, \end{cases} \tag{3.16}$$

for $\mathbf{i}_{(l)} \in \mathcal{S}_{(l)}$ and $\mathbf{i}_{(l+1)} \in \mathcal{S}_{(l+1)}$,

$$\pi_{(m,l+1)} = \mathbf{x}_{(m,l)} R_{(l)} \quad \text{and} \quad \tilde{Q}_{(m,l+1)} = P_{\mathbf{x}_{(m,l)}} \tilde{Q}_{(m,l)} R_{(l)}. \tag{3.17}$$

At iteration m, the recursion ends and backtracking starts when $\tilde{Q}_{(m,l+1)}$ in (3.17) is the last aggregated CTMC and solved exactly to give $T_{(m,l+1)} = \mathbf{e}\pi_{(m+1,l+1)}$, where $\pi_{(m+1,l+1)} \tilde{Q}_{(m,l+1)} = \mathbf{0}$ and $\pi_{(m+1,l+1)}\mathbf{e} = 1$. The level to end recursion depends on available memory since there must be space to store and factorize the aggregated CTMC at that level. When the initial approximation is positive (i.e., $\pi_{(0,0)} > \mathbf{0}$), the aggregated CTMCs $\tilde{Q}_{(m,l+1)}$ are irreducible [28], and the ML method has been observed to converge if a sufficient number of smoothings are performed to improve the approximate solution vector, $\pi_{(m,l)}$, at each level.

We remark that in contrast to block iterative methods, the ML iteration matrix in (3.13) changes from iteration to iteration, and hence, the method is nonstationary. Nevertheless, the *aggregation* (or *restriction*) operator $R_{(l)} \in \mathbb{R}_{\geq 0}^{|\mathcal{S}_{(l)}| \times |\mathcal{S}_{(l+1)}|}$ in (3.15) is constant and need not be stored since it is defined by $f_{(l)}$. At level l, the $|\mathcal{S}_{(l)}| = \prod_{h=l+1}^{H} n_l$ states represented by $(H - l)$-tuples are mapped to the $|\mathcal{S}_{(l+1)}| = \prod_{h=l+2}^{H} n_l$ states represented by $(H - l - 1)$-tuples by aggregating the leading dimension $\mathcal{S}^{(l+1)}$ in $\mathcal{S}_{(l)}$. This would correspond to an aggregation based on a contiguous and noninterleaved block partitioning if the states in $\mathcal{S}_{(l)}$ were ordered antilexicographically. On the other hand, the *disaggregation* (or *prolongation*) operator $P_{\mathbf{x}_{(m,l)}} \in \mathbb{R}^{|\mathcal{S}_{(l+1)}| \times |\mathcal{S}_{(l)}|}$ in (3.16) depends on the smoothed vector $\mathbf{x}_{(m,l)}$ in (3.14) and has the nonzero structure of $R_{(l)}^T$. Therefore, $P_{\mathbf{x}_{(m,l)}}$ can be stored in a vector of length $|\mathcal{S}_{(l)}|$ since it has one nonzero per column by definition. These vectors amount to a total storage of $\sum_{l=0}^{H-1} \prod_{h=l+1}^{H} n_h$ floating-point values if recursion terminates at level H.

In [28], it is shown that $\tilde{Q}_{(m,l+1)}$ can be expressed as a sum of Kronecker products using at most K vectors of length $|\mathcal{S}_{(l+1)}|$ and matrices corresponding to the factors $(l+2)$ through H. More specifically, the $\mathbf{i}_{(l+1)}$st element of the vector corresponding to the kth term in the Kronecker representation at level $(l + 1)$ for iteration m, $\mathbf{i}_{(l+1)} \in \mathcal{S}_{(l+1)}$, and $k = 1, \ldots, K$ is defined as

$$\mathbf{a}_{(m,l+1),k}(\mathbf{i}_{(l+1)})$$

$$= \frac{\sum_{\mathbf{j}_{(l)} \in \mathcal{S}_{(l)} : f_{(l)}(\mathbf{j}_{(l)}) = \mathbf{i}_{(l+1)}} \mathbf{x}_{(m,l)}(\mathbf{j}_{(l)}) \, \mathbf{a}_{(m,l),k}(\mathbf{j}_{(l)}) \left(\mathbf{e}_{(\mathbf{j}_{(l)})_{l+1}}^T Q_k^{(l+1)} \mathbf{e} \right)}{\pi_{(m,l+1)}(\mathbf{i}_{(l+1)})}, \tag{3.18}$$

where $\mathbf{a}_{(m,0),k} = \mathbf{e}$, $(\mathbf{j}_{(l)})_{l+1} \in \mathcal{S}^{(l+1)}$, and $\mathbf{e}_{(\mathbf{j}_{(l)})_{l+1}}$ is the $(\mathbf{j}_{(l)})_{l+1}$st column of I. Then

$$
\tilde{Q}_{(m,l+1)} = \sum_{k=1}^{K} \mathrm{diag}(\mathbf{a}_{(m,l+1),k}) \bigotimes_{h=l+2}^{H} Q_k^{(h)}
$$
$$
- \sum_{k=1}^{K} \mathrm{diag}(\mathbf{a}_{(m,l+1),k}) \bigotimes_{h=l+2}^{H} \mathrm{diag}\left(Q_k^{(h)}\mathbf{e}\right). \tag{3.19}
$$

The second summation in (3.19) returns a diagonal matrix, which sums the rows of $\tilde{Q}_{(m,l+1)}$ to zero. Furthermore, the vectors $\mathbf{a}_{(m,0),k}$ for $k = 1,\ldots,K$ at level 0 consist of all 1s and therefore need not be stored. If the recursion ends at level H, then $\tilde{Q}_{(m,H)}$ is a (1×1) CTMC equal to zero and need not be stored since its steady-state vector is (1). Note that $\mathbf{a}_{(m,l+1),k} = \mathbf{e}$ for those k that have either a single $Q_k^{(h)} \neq I$ for $h = 1,\ldots,H$ or all $Q_k^{(h)} = I$ for $h = l+2,\ldots,H$. Such vectors need not be stored either. The K vectors at a particular level have the same length but vary in length from $\prod_{h=2}^{H} n_h$ at level 1 to n_H at level $(H-1)$, implying a storage requirement of at most $K \sum_{l=1}^{H-1} \prod_{h=l+1}^{H} n_h$ floating-point values to facilitate the Kronecker representation of the aggregated CTMCs. The grouping of factors will further reduce the storage requirement for vectors.

Example 1 (ctnd.). Consider our three-dimensional problem with parameter set $(\lambda_1, \lambda_2, \lambda_3, \mu_1, \mu_2, \mu_3, \mu) = (1, 2, 3, 2, 4, 6, 10)$, initial distribution $\pi_{(0,0)} = \mathbf{e}/12$, $\alpha_{(0,0)} = 22$, and $\eta_{(0,0)} = \nu_{(0,0)} = 1$. Since

$\tilde{Q}_{(0,0)} =$

	000	001	010	011	020	021	100	101	110	111	120	121
000	−6	3	2				1					
001	6	−9		2				1				
010	4		−10	3	2				1			
011		4	6	−13		2				1		
020			4		−8	3					1	
021				4	6	−11						1
100	2						−7	3	2			
101		2					6	−10		2		
110			2				4		−11	3	2	
111				2				4	6	−14		2
120					2				4		−9	3
121	10					2				4	6	−22

$\mathbf{x}_{(0,0)} = \pi_{(0,0)}(I + \tilde{Q}_{(0,0)}/22)$ from (3.14) yields

$$\mathbf{x}_{(0,0)} = \left(\frac{19}{132}, \frac{11}{132}, \frac{13}{132}, \frac{10}{132}, \frac{12}{132}, \frac{9}{132}, \frac{13}{132}, \frac{10}{132}, \frac{12}{132}, \frac{9}{132}, \frac{11}{132}, \frac{3}{132} \right).$$

Furthermore,

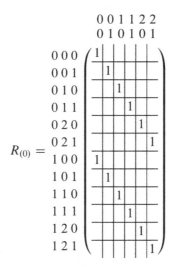

and

from (3.15) and (3.16), respectively. Hence, the 12 states represented by three-tuples in $\mathcal{S}_{(0)} = \mathcal{S}$ are mapped to the 6 states represented by two-tuples in $\mathcal{S}_{(1)}$. For instance, states $(0,0,0)$ and $(1,0,0)$ are mapped to $(0,0)$, whereas states $(0,0,1)$ and $(1,0,1)$ are mapped to $(0,1)$. Using $R_{(0)}$ in the first part of (3.17), we obtain the starting approximation at level 1 as

$$\boldsymbol{\pi}_{(0,1)} = \left(\frac{32}{132}, \frac{21}{132}, \frac{25}{132}, \frac{19}{132}, \frac{23}{132}, \frac{12}{132} \right).$$

Through (3.18), the four vectors used to represent the aggregated CTMC at level 1 are computed as

$$\mathbf{a}_{(0,1),1} = \left(\frac{45}{32}, \frac{31}{21}, \frac{37}{25}, \frac{28}{19}, \frac{34}{23}, \frac{5}{4}\right),$$

$$\mathbf{a}_{(0,1),2} = \mathbf{a}_{(0,1),3} = \mathbf{e},$$

$$\mathbf{a}_{(0,1),4} = \left(\frac{65}{16}, \frac{100}{21}, \frac{24}{5}, \frac{90}{19}, \frac{110}{23}, \frac{5}{2}\right),$$

and the aggregated CTMC is expressed as

$$\tilde{Q}_{(0,1)} = P_{\mathbf{x}_{(0,0)}} \tilde{Q}_{(0,0)} R_{(0)}$$

$$= \sum_{k=1}^{4} \text{diag}(\mathbf{a}_{(0,1),k}) \bigotimes_{h=2}^{4} Q_k^{(h)} - \sum_{k=1}^{4} \text{diag}(\mathbf{a}_{(0,1),k}) \bigotimes_{h=2}^{4} \text{diag}\left(Q_k^{(h)}\mathbf{e}\right).$$

Observe that the effect of $\mathbf{a}_{(0,1),1}$ in the first term of the first summation is to the diagonal of $\tilde{Q}_{0,1}$ since $Q_1^{(2)} = Q_1^{(3)} = I$. But this effect is canceled by the first term of the second summation simply because $\text{diag}(Q_1^{(2)}\mathbf{e}) = \text{diag}(Q_1^{(3)}\mathbf{e}) = I$. Hence, we may very well set $\mathbf{a}_{(0,1),1} = \mathbf{e}$ as suggested previously. Furthermore, $\mathbf{a}_{(0,1),2} = \mathbf{a}_{(0,1),3} = \mathbf{e}$ since $Q_2^{(1)} = Q_3^{(1)} = I$ and $\mathbf{a}_{(0,0),k} = \mathbf{e}$ for $k = 1,\ldots,4$. Therefore, we implicitly have

$$
\tilde{Q}_{(0,1)} =
\begin{array}{c}
\\
\\
0\,0 \\
0\,1 \\
1\,0 \\
1\,1 \\
2\,0 \\
2\,1
\end{array}
\begin{array}{cccccc}
0 & 0 & 1 & 1 & 2 & 2 \\
0 & 1 & 0 & 1 & 0 & 1 \\
\left(\begin{array}{cc|cc|cc}
-5 & 3 & 2 & & & \\
6 & -8 & & 2 & & \\
\hline
4 & & -9 & 3 & 2 & \\
& 4 & 6 & -12 & & 2 \\
\hline
& & 4 & & -7 & 3 \\
\frac{5}{2} & & & 4 & 6 & -\frac{25}{2}
\end{array}\right)
\end{array}.
$$

In the next step, similar operations will be carried out at level 1 unless the aggregated CTMC is solved exactly, upon which backtracking from recursion starts for iteration m.

The ML method we discussed follows a *V-cycle* [96] at each iteration. That is, starting from the finest level, at each step it smooths the current approximation and moves to a coarser level by aggregation until it reaches a level at which the aggregated CTMC can be solved exactly. Once the exact solution is obtained at the coarsest level, the method starts moving in the opposite direction. At each step on the way to the finest level, the method disaggregates the current approximation passed by the coarser level and smooths it. Furthermore, the state spaces $\mathcal{S}^{(h)}$ are aggregated according to the *fixed* ordering $h = 1,\ldots,H$. However, in contrast to the ML method for sparse MCs in [92], the definition of the aggregated state spaces follows naturally from the Kronecker representation in (2.1), and the aggregated CTMCs can also be represented using Kronecker products as shown in (3.19). In [33], a sophisticated class of ML methods is given. The methods there are capable of using JOR and SOR as smoothers, performing the *W-* and *F-cycles* inspired by multigrid [17, 127, 150], and aggregating the state spaces according to *circular* and *adaptive* orderings. Here, W-cycle refers to invoking at each level two recursive calls to the next coarser level, whereas an F-cycle at a level can be viewed as a recursive call to a W-cycle followed by a recursive call to a V-cycle at the next coarser level. In the circular ordering of aggregation, at the beginning of each ML cycle at the finest level a circular shift of subsystem indices is performed to achieve fairness in aggregating subsystem state spaces. Hence, every H cycles, each subsystem will have received the opportunity to get aggregated first. On the other hand, in the adaptive ordering of aggregation, subsystem indices are sorted according to the residual norms restricted to the corresponding subsystem state space at the end of the ML cycle, and aggregation of subsystems in this sorted order in the next cycle is performed. This ensures that subsystems that have smaller residual norms are aggregated earlier at finer levels since small residual norms are expected to be indicative of good approximations in those subsystems. Numerical experiments in [33] prove ML methods to be very strong, robust, and scalable solvers for MCs based on Kronecker products.

The convergence properties of the class of ML methods in [33] are discussed in [35]; however, it is not clear how its behavior would be affected if block iterative methods, such as BJOR and BSOR, were used as smoothers rather than power, JOR, and SOR. Note that BJOR and BSOR should normally not use a direct method for the solution of the diagonal blocks when employed as smoothers with the ML method since the aggregated CTMC at each level changes from iteration to iteration and the factorization may be too time consuming to offset. In [86], an efficient algorithm that finds a *nearly completely decomposable* (NCD) [53, 59, 110, 139] partitioning of S in the presence of functional transitions for a user-specified decomposability parameter is given. Since IAD using NCD partitionings has a certain rate of convergence guarantees [138] (cf. [105, 106, 125]), the algorithm may be useful in the context of ML methods to determine the loosely coupled dimensions to be aggregated first in a given iteration.

Chapter 4
Decompositional Methods

Among the iterative methods discussed in the previous chapter, ML methods perform better on a larger number of problems in the literature [33, 35]. However, there are certain classes of problems for which other methods could be preferred. The first such method we present in this chapter is iterative and based on *decomposing* a system into its subsystems, analyzing the subsystems individually for their steady state, and putting back the individual solutions together using disaggregation in a correction step [4]. This method is able to compute the steady-state solution exactly up to computer precision to a system having weakly interacting subsystems in a relatively small number of iterations and with modest memory requirements. There are also iterative methods based on polyhedra theory [46] and disaggregation, such as that in [30] for SANs, which provides satisfactory lower and upper bounds on the solution only if the interactions among subsystems are weak or the rates of synchronized transitions are more or less independent of the states of subsystems.

4.1 Exact Versus Approximative

Consider the separation of Q_O in (2.1) into two terms as in

$$Q_O = Q_{\text{local}} + Q_{\text{synchronized}},$$

where Q_{local} and $Q_{\text{synchronized}}$ correspond to those parts of Q_O associated with local and synchronized transitions, respectively. Without loss of generality, we adopt the enumeration of the K terms as in Example 1 and let the first H represent local transitions, the hth being associated with subsystem h for $h = 1, \ldots, H$. The remaining $(K - H)$ terms correspond to synchronized transitions. Hence,

$$Q_{\text{local}} = \sum_{k=1}^{H} \bigotimes_{h=1}^{H} Q_k^{(h)} \quad \text{and} \quad Q_{\text{synchronized}} = \sum_{k=H+1}^{K} \bigotimes_{h=1}^{H} Q_k^{(h)}.$$

T. Dayar, *Analyzing Markov Chains using Kronecker Products: Theory and Applications*, SpringerBriefs in Mathematics, DOI 10.1007/978-1-4614-4190-8_4, © Tuğrul Dayar 2012

Recall that this enumeration necessarily implies that $Q_k^{(h)} = I_{n_h}$ for $h = 1, \ldots, H$ and $k \neq h$ due to the definition of local transitions. Furthermore, observe that the irreducibility of Q does not imply the irreducibility of the local transition rate matrices $Q_h^{(h)}$ for $h = 1, \ldots, H$.

Now, let $Q = U - L$ be the forward GS splitting of the infinitesimal generator in Kronecker form, where $U = Q_D + Q_{U(H)}$ corresponds to its upper-triangular part and $L = Q_{L(H)}$ contains its negated strictly lower-triangular part as in (3.7). Furthermore, let the aggregation operator $R^{(h)} \in \mathbb{R}_{\geq 0}^{n \times n_h}$ [cf. (3.15)] for $h = 1, \ldots, H$ be associated with the mapping $f_{(h)} : S \longrightarrow S^{(h)}$ and have its $(\mathbf{s}, s_h)$th entry be given by

$$r^{(h)}(\mathbf{s}, s_h) = \begin{cases} 1 & \text{if } f_{(h)}(\mathbf{s}) = s_h \\ 0 & \text{otherwise} \end{cases} \quad \text{for } \mathbf{s} \in S \text{ and } s_h \in S^{(h)}.$$

Observe that the mapping $f_{(h)}$ represents the aggregation of all dimensions except the hth. In Kronecker form,

$$R^{(h)} = \left(\bigotimes_{l=1}^{h-1} I_{n_l} \mathbf{e} \right) \otimes I_{n_h} \otimes \left(\bigotimes_{l=h+1}^{H} I_{n_l} \mathbf{e} \right) \quad \text{for } h = 1, \ldots, H.$$

On the other hand, let the disaggregation operator $P_{\pi_{(m)}}^{(h)} \in \mathbb{R}^{n_h \times n}$ [cf. (3.16)] for $h = 1, \ldots, H$ be associated with the mapping $f_{(h)}$ and have its $(s_h, \mathbf{s})$th entry be given by

$$P_{\pi_{(m)}}^{(h)}(s_h, \mathbf{s}) = \begin{cases} \frac{\pi_{(m)}(\mathbf{s})}{\pi_{(m)}^{(h)}(s_h)} & \text{if } f_{(h)}(\mathbf{s}) = s_h \\ 0 & \text{otherwise} \end{cases} \quad \text{for } \mathbf{s} \in S \text{ and } s_h \in S^{(h)},$$

where $\pi_{(m)}^{(h)} = \pi_{(m)} R^{(h)}$. Then the decompositional iterative method can be stated [4] for a user-specified stopping tolerance, tol, as in Algorithm 3.

The algorithm starts by initializing the correction vector, $\mathbf{y}_{(m)}$, to zero and the solution vector, $\pi_{(m)}$, to the uniform distribution. Then each system of local equations is solved subject to a normalization condition. If $Q_h^{(h)}$ is irreducible, then a unique new local solution vector $\pi_{(m+1)}^{(h)} > 0$ can be computed. This is so because each system to be solved has a zero-sum right-hand side vector, $\mathbf{v}^{(h)}(\pi_{(m)})$ [i.e., $\mathbf{v}^{(h)}(\pi_{(m)})\mathbf{e} = 0$] due to the particular way in which synchronized transition rate matrices, $Q_k^{(h)}$ for $k = H + 1, \ldots, K$, are specified. On the other hand, when $Q_h^{(h)}$ is reducible, we consider a homogeneous system in which the aggregated matrix $Q^{(h)}(\pi_{(m)})$ is used. The aggregated matrix is irreducible if Q is irreducible and $\pi_{(m)} > 0$. Hence, the existence of a unique $\pi_{(m+1)}^{(h)} > 0$ is also guaranteed in this case. Since $Q_k^{(h)}$ for $k = H + 1, \ldots, K$ are in general very sparse, the enumeration process associated with the nonzeros in $Q_{\text{synchronized}}$ to form $\mathbf{v}^{(k)}(\boldsymbol{\pi}_{(m)})$, or $Q^{(h)}(\boldsymbol{\pi}_{(m)})$, can be handled systematically. There are differences from a computational point of view between the two alternative solution steps. In the former case, $Q_h^{(h)}$ is constant and already available in sparse format; the right-hand-side vector is dependent on $\boldsymbol{\pi}_{(m)}$. In the latter case, $Q^{(h)}(\boldsymbol{\pi}_{(m)})$ needs to be reconstructed at each iteration, and it is the coefficient matrix that is dependent on $\boldsymbol{\pi}_{(m)}$. Hence, the two approaches to obtaining $\boldsymbol{\pi}_{(m+1)}^{(h)}$ are not equivalent except at steady state [i.e., $\boldsymbol{\pi}_{(m)} = \boldsymbol{\pi}_{(m+1)} = \boldsymbol{\pi}$].

In the next step, the new correction vector, $\mathbf{y}_{(m+1)}$, is obtained by solving an upper-triangular system in Kronecker form. Since Q is assumed to be irreducible, $\mathbf{y}_{(m+1)}$ is computed through a GS relaxation on Q with a zero sum but nonzero right-hand side.

The last step subtracts $\mathbf{y}_{(m+1)}$ from the Kronecker product of $\boldsymbol{\pi}_{(m+1)}^{(h)}$ for $h = 1, \ldots, H$ to form the new solution vector, $\boldsymbol{\pi}_{(m+1)}$, and then the iteration number is incremented. These steps are repeated until the residual infinity norm becomes smaller than tol.

Algorithm 3. *Decompositional iterative method with GS correction step*

$m = 0; \mathbf{y}_{(m)} = \mathbf{0}; \boldsymbol{\pi}_{(m)} = \mathbf{e}^T/n;$
Repeat
    For $h = 1$ to H,
        If $Q_h^{(h)}$ is irreducible, solve $\boldsymbol{\pi}_{(m+1)}^{(h)} Q_h^{(h)} = \mathbf{v}^{(k)}(\boldsymbol{\pi}_{(m)})$,
            where $\mathbf{v}^{(k)}(\boldsymbol{\pi}_{(m)}) = -\boldsymbol{\pi}_{(m)} Q_{\text{synchronized}} R^{(h)}$,
        Else solve $\boldsymbol{\pi}_{(m+1)}^{(h)} Q^{(h)}(\boldsymbol{\pi}_{(m)}) = \mathbf{0}$,
            where $Q^{(h)}(\boldsymbol{\pi}_{(m)}) = Q_h^{(h)} + P_{\boldsymbol{\pi}_{(m)}}^{(h)} Q_{\text{synchronized}} R^{(h)}$,
        subject to $\boldsymbol{\pi}_{(m+1)}^{(h)} \mathbf{e} = 1$;
    Solve $\mathbf{y}_{(m+1)} U = \mathbf{y}_{(m)} L + \left(\bigotimes_{h=1}^{H} \boldsymbol{\pi}_{(m+1)}^{(h)} \right) Q$;
    $\boldsymbol{\pi}_{(m+1)} = \left(\bigotimes_{h=1}^{H} \boldsymbol{\pi}_{(m+1)}^{(h)} \right) - \mathbf{y}_{(m+1)}$ subject to $\boldsymbol{\pi}_{(m+1)} \mathbf{e} = 1$;
    $m = m + 1$;
Until $\| \boldsymbol{\pi}_{(m)} Q \|_\infty < tol$.

Algorithm 3 is coded into the APNN toolbox [7], and numerical experiments are carried out [4] on larger and slightly different versions of Example 1.

Example 1 (ctnd.). We consider a fail-repair system with five subsystems each having 20 states (i.e., $H = 5$ and $n_h = 20$ for $h = 1, \ldots, H$) to yield a CTMC in Kronecker form of $n = 3,200,000$ states. We let local failure and repair rates take the values $(\lambda_1, \lambda_2, \lambda_3, \lambda_4, \lambda_5) = (0.4, 0.5, 0.6, 0.7, 0.8)$ and $(\mu_1, \mu_2, \mu_3, \mu_4, \mu_5) = (0.3, 0.4, 0.5, 0.6, 0.7)$, respectively. Furthermore, we introduce four synchronized transitions (i.e., $K = H + 4$), each having a rate of 0.005. The first synchronized transition takes the system to state $(0, 0, 0, 0, 0)$ from state $(4, 4, 4, 4, 4)$, the second takes the system to state $(5, 5, 5, 5, 5)$ from state $(9, 9, 9, 9, 9)$, the third takes the system to state $(10, 10, 10, 10, 10)$ from state $(14, 14, 14, 14, 14)$, and the fourth takes the system to state $(15, 15, 15, 15, 15)$ from state $(19, 19, 19, 19, 19)$. These transitions can be considered as batch repairs of four failed redundant components in each subsystem when the system is in any one of the states $(4, 4, 4, 4, 4)$, $(9, 9, 9, 9, 9)$, $(14, 14, 14, 14, 14)$, and $(19, 19, 19, 19, 19)$.

Experiments are performed on a PC with an Intel Core2 Duo 1.83-GHz processor having 4 GB of main memory. The decompositional method is compared with the solvers Jacobi, GS, BGS, GMRES with a Krylov subspace size of 20, BICGSTAB, transpose-free quasiminimal residual (TFQMR), BGS-preconditioned GMRES(20), TFQMR, BICGSTAB, and ML with one pre- and one postsmoothing using GS, W-cycle, and circular order of aggregating subsystems in each cycle. The solvers are compared in terms of the number of iterations to converge to a tolerance of $tol = 10^{-8}$ on the infinity norm of the residual, elapsed CPU time, and amount of allocated main memory. The stopping test is executed every ten iterations in the decompositional solver as in the Jacobi, GS, and BGS solvers. The diagonal blocks associated with the BGS solver and the BGS preconditioner for projection methods at level 3 are LU factorized [32] using column approximate minimum degree (COLAMD) ordering [51]. The number of nonzeros generated during the LU factorization of the 8,000 diagonal blocks of order 400 is accounted for in the memory consumed by solvers utilizing BGS.

It is observed that convergence becomes very fast for the decompositional solver when the synchronized transition rates are small since the subsystems in that case are nearly independent and the Kronecker product of the local solutions yields a very good approximation to the solution early in the iteration. The solver corresponding to Algorithm 3 requires 171 MB and merely 10 iterations to converge to the solution in 25 s. The second best solver is ML, which takes 232 MB and 24 iterations to converge to the solution in 99 s. It improves only slightly as the synchronized transition rates become smaller. The only other competitive solver is BICGSTAB, which takes 143 iterations and 127 s, requiring 244 MB. In particular, GMRES(20), BICGSTAB, and TFQMR do not benefit from BGS preconditioning. The performances of the Jacobi, GS, and BGS solvers are not affected by a change in the rates of synchronized transitions. BGS performs very poorly due to the large time per iteration. BGS and BGS-preconditioned projection methods require considerably more memory than the other methods because of the need to store factors of diagonal blocks and, in the latter case, also a larger number of vectors. For instance, the space taken by BGS-preconditioned BICGSTAB is 1,010 MB. Memorywise, the decompositional solver requires about 1.5 times that of Jacobi and GS, but less than ML, and therefore can be considered to be memory efficient.

The scalability of the decompositional solver is investigated for an increasing number of subsystems when the four synchronized transition rates are relatively small compared to those in local transition rate matrices. It is observed that the number of iterations to converge decreases as the number of subsystems increases. This is due to the decrease in the throughputs of synchronized transitions for a larger number of subsystems (because the steady-state probabilities of states in which synchronized transitions can take place become smaller), leading to more independent subsystems. This is different from the behavior of the ML method, which takes more or less the same number of iterations to converge as the number of subsystems increases. The scalability of the decompositional solver is also investigated for an increasing number of synchronized transitions when the number of subsystems is kept at five and the rates of synchronized transitions are relatively small compared to those in the local transition rate matrices. As expected, the results indicate that the time the decompositional solver takes to converge is affected linearly by an increase in the number of synchronized transitions.

Another class of iterative methods consists of those that are approximative. For instance, the method in [31] for superposed GSPNs operates at a fine level only on states having higher steady-state probabilities; the remaining states are aggregated and treated at a coarse level. The steady-state vector can be stored with significant savings due to its compact representation as a Kronecker product of the aggregated subsystem steady-state vectors. The second class of iterative methods presented in this chapter are also approximative. They are to be employed for closed queueing networks with phase-type service distributions and arbitrary buffer sizes when a few digits of accuracy in the computed solution is sufficient for analysis purposes [56]. We discuss two such methods. Each decomposes the network into subnetworks. They differ in the way the decompositions are obtained and the solutions to subnetworks are put together. These approximative methods require the modeling of subnetworks whose product state-space sizes are larger than their reachable state-space sizes. This motivates the discussion on how the reachability problem is handled in such models.

4.2 Handling Unreachable States

So far, we have assumed that the state space, S, is equal to the product state space, $\times_{h=1}^{H} S^{(h)}$, and has no unreachable states. We now relax this assumption and discuss the case of unreachable states.

Example 2. Consider an irreducible CTMC with five states corresponding to two interacting subsystems. Hence, $|S| = 5$ and $H = 2$. Now, note that unless $|S^{(1)}| = 1$ and $|S^{(2)}| = 5$ or vice versa, the size of $S^{(1)} \times S^{(2)}$ will not be equal to that of S. Hence, we assume that one of the subsystems has more than one state; without loss of generality, we let $1 < |S^{(1)}| \leq |S^{(2)}| \leq 5$ and investigate the possibilities. For simplicity, we do not consider *isomorphic* cases throughout the discussion and let N denote the resulting number of partitions in the representation of the reachable state space. By isomorphic cases we mean identical cases up to a renumbering of the states.

i. $|\mathcal{S}^{(1)}| = |\mathcal{S}^{(2)}| = 5$, $\mathcal{S}^{(1)}$ and $\mathcal{S}^{(2)}$ are partitioned into $N = 5$ subsets of one state each. That is, $\mathcal{S}_p^{(h)} = \{p - 1\}$ for $p = 1, \ldots, N$ and $\mathcal{S}^{(h)} = \bigcup_{p=1}^{N} \mathcal{S}_p^{(h)}$ for $h = 1, 2$. Then $|\mathcal{S}^{(1)} \times \mathcal{S}^{(2)}| = 25$, but

$$\mathcal{S} = \bigcup_{p=1}^{N} \mathcal{S}_p^{(1)} \times \mathcal{S}_p^{(2)} = \{(0,0)\} \cup \{(1,1)\} \cup \{(2,2)\} \cup \{(3,3)\} \cup \{(4,4)\}.$$

ii. $|\mathcal{S}^{(1)}| = 4$, $|\mathcal{S}^{(2)}| = 5$, $\mathcal{S}^{(1)}$ and $\mathcal{S}^{(2)}$ are partitioned into $N = 4$ subsets, $\mathcal{S}^{(2)}$ having a subset of size 2. That is, $\mathcal{S}_p^{(h)} = \{p - 1\}$ for $p = 1, 2, 3$, $\mathcal{S}_4^{(1)} = \{3\}$, $\mathcal{S}_4^{(2)} = \{3, 4\}$, and $\mathcal{S}^{(h)} = \bigcup_{p=1}^{N} \mathcal{S}_p^{(h)}$ for $h = 1, 2$. Then $|\mathcal{S}^{(1)} \times \mathcal{S}^{(2)}| = 20$, but

$$\mathcal{S} = \bigcup_{p=1}^{N} \mathcal{S}_p^{(1)} \times \mathcal{S}_p^{(2)} = \{(0,0)\} \cup \{(1,1)\} \cup \{(2,2)\} \cup \{(3,3), (3,4)\}.$$

iii. $|\mathcal{S}^{(1)}| = 3$, $|\mathcal{S}^{(2)}| = 5$, $\mathcal{S}^{(1)}$ and $\mathcal{S}^{(2)}$ are partitioned into $N = 3$ subsets, $\mathcal{S}^{(2)}$ having two subsets of size 2. That is, $\mathcal{S}_1^{(1)} = \mathcal{S}_1^{(2)} = \{0\}$, $\mathcal{S}_2^{(1)} = \{1\}$, $\mathcal{S}_2^{(2)} = \{1, 2\}$, $\mathcal{S}_3^{(1)} = \{2\}$, $\mathcal{S}_3^{(2)} = \{3, 4\}$, and $\mathcal{S}^{(h)} = \bigcup_{p=1}^{N} \mathcal{S}_p^{(h)}$ for $h = 1, 2$. Then $|\mathcal{S}^{(1)} \times \mathcal{S}^{(2)}| = 15$, but

$$\mathcal{S} = \bigcup_{p=1}^{N} \mathcal{S}_p^{(1)} \times \mathcal{S}_p^{(2)} = \{(0,0)\} \cup \{(1,1), (1,2)\} \cup \{(2,3), (2,4)\}.$$

iv. $|\mathcal{S}^{(1)}| = 3$, $|\mathcal{S}^{(2)}| = 5$, $\mathcal{S}^{(1)}$ and $\mathcal{S}^{(2)}$ are partitioned into $N = 3$ subsets, $\mathcal{S}^{(2)}$ having a subset of size 3. That is, $\mathcal{S}_1^{(1)} = \mathcal{S}_1^{(2)} = \{0\}$, $\mathcal{S}_2^{(1)} = \{1\}$, $\mathcal{S}_2^{(2)} = \{1\}$, $\mathcal{S}_3^{(1)} = \{2\}$, $\mathcal{S}_3^{(2)} = \{2, 3, 4\}$, and $\mathcal{S}^{(h)} = \bigcup_{p=1}^{N} \mathcal{S}_p^{(h)}$ for $h = 1, 2$. Then $|\mathcal{S}^{(1)} \times \mathcal{S}^{(2)}| = 15$, but

$$\mathcal{S} = \bigcup_{p=1}^{N} \mathcal{S}_p^{(1)} \times \mathcal{S}_p^{(2)} = \{(0,0)\} \cup \{(1,1)\} \cup \{(2,2), (2,3), (2,4)\}.$$

v. $|\mathcal{S}^{(1)}| = 2$, $|\mathcal{S}^{(2)}| = 5$, $\mathcal{S}^{(1)}$ and $\mathcal{S}^{(2)}$ are partitioned into $N = 2$ subsets, $\mathcal{S}^{(2)}$ having a subset of size 4. That is, $\mathcal{S}_1^{(1)} = \{0\}$, $\mathcal{S}_1^{(2)} = \{0\}$, $\mathcal{S}_2^{(1)} = \{1\}$, $\mathcal{S}_2^{(2)} = \{1, 2, 3, 4\}$, and $\mathcal{S}^{(h)} = \bigcup_{p=1}^{N} \mathcal{S}_p^{(h)}$ for $h = 1, 2$. Then $|\mathcal{S}^{(1)} \times \mathcal{S}^{(2)}| = 10$, but

$$\mathcal{S} = \bigcup_{p=1}^{N} \mathcal{S}_p^{(1)} \times \mathcal{S}_p^{(2)} = \{(0,0)\} \cup \{(1,1), (1,2), (1,3), (1,4)\}.$$

vi. $|\mathcal{S}^{(1)}| = 4$, $|\mathcal{S}^{(2)}| = 4$, $\mathcal{S}^{(1)}$ and $\mathcal{S}^{(2)}$ are partitioned into $N = 3$ subsets, $\mathcal{S}^{(1)}$ and $\mathcal{S}^{(2)}$ each having a subset of size 2. That is, $\mathcal{S}_1^{(1)} = \mathcal{S}_1^{(2)} = \{0\}$, $\mathcal{S}_2^{(1)} = \{1\}$, $\mathcal{S}_2^{(2)} = \{1, 2\}$, $\mathcal{S}_3^{(1)} = \{2, 3\}$, $\mathcal{S}_3^{(2)} = \{3\}$, and $\mathcal{S}^{(h)} = \bigcup_{p=1}^{N} \mathcal{S}_p^{(h)}$ for $h = 1, 2$. Then $|\mathcal{S}^{(1)} \times \mathcal{S}^{(2)}| = 16$, but

$$\mathcal{S} = \bigcup_{p=1}^{N} \mathcal{S}_p^{(1)} \times \mathcal{S}_p^{(2)} = \{(0, 0)\} \cup \{(1, 1), (1, 2)\} \cup \{(2, 3), (3, 3)\}.$$

vii. $|\mathcal{S}^{(1)}| = 3$, $|\mathcal{S}^{(2)}| = 4$, $\mathcal{S}^{(1)}$ and $\mathcal{S}^{(2)}$ are partitioned into $N = 2$ subsets, $\mathcal{S}^{(1)}$ having a subset of size 2 and $\mathcal{S}^{(2)}$ having a subset of size 3. That is, $\mathcal{S}_1^{(1)} = \{0\}$, $\mathcal{S}_1^{(2)} = \{0, 1, 2\}$, $\mathcal{S}_2^{(1)} = \{1, 2\}$, $\mathcal{S}_2^{(2)} = \{3\}$, and $\mathcal{S}^{(h)} = \bigcup_{p=1}^{N} \mathcal{S}_p^{(h)}$ for $h = 1, 2$. Then $|\mathcal{S}^{(1)} \times \mathcal{S}^{(2)}| = 12$, but

$$\mathcal{S} = \bigcup_{p=1}^{N} \mathcal{S}_p^{(1)} \times \mathcal{S}_p^{(2)} = \{(0, 0), (0, 1), (0, 2)\} \cup \{(1, 3), (2, 3)\}.$$

viii. $|\mathcal{S}^{(1)}| = 3$, $|\mathcal{S}^{(2)}| = 3$, $\mathcal{S}^{(1)}$ and $\mathcal{S}^{(2)}$ are partitioned into $N = 2$ subsets, $\mathcal{S}^{(1)}$ and $\mathcal{S}^{(2)}$ each having a subset of size 2. That is, $\mathcal{S}_1^{(1)} = \{0\}$, $\mathcal{S}_1^{(2)} = \{0\}$, $\mathcal{S}_2^{(1)} = \{1, 2\}$, $\mathcal{S}_2^{(2)} = \{1, 2\}$, and $\mathcal{S}^{(h)} = \bigcup_{p=1}^{N} \mathcal{S}_p^{(h)}$ for $h = 1, 2$. Then $|\mathcal{S}^{(1)} \times \mathcal{S}^{(2)}| = 9$, but

$$\mathcal{S} = \bigcup_{p=1}^{N} \mathcal{S}_p^{(1)} \times \mathcal{S}_p^{(2)} = \{(0, 0)\} \cup \{(1, 1), (1, 2), (2, 1), (2, 2)\}.$$

As observed, excluding the isomorphic cases there are eight different ways in which one can obtain $|\mathcal{S}| = 5$ from a two-dimensional product state space using $2 \leq N \leq 5$ partitions. However, there are certain state-space sizes that never lead to $|\mathcal{S}| = 5$ no matter how the state spaces are partitioned. Consider $|\mathcal{S}^{(1)}| = 2$ and $|\mathcal{S}^{(2)}| = 4$, for instance.

We see in the following section and next chapter that given the state-space size, $\mathcal{S}^{(h)}$, of each subsystem for $h = 1, \ldots, H$, the number of possibilities available in choosing the number of partitions, N, is large enough to accommodate a representation of the state space without unreachable states. Once N is determined, the CTMC is represented as an $(N \times N)$ block matrix, with block (p, w) being defined as a sum of Kronecker products capturing the transitions from states in $\times_{h=1}^{H} \mathcal{S}_p^{(h)}$ to states in $\times_{h=1}^{H} \mathcal{S}_w^{(h)}$ for $p, w = 1, \ldots, N$. This is essentially the idea behind HMMs [19]. All iterative methods discussed in Chap. 3 are implemented in the APNN toolbox [7] for an $(N \times N)$ block matrix whose blocks are sums of Kronecker products. The off-diagonal blocks of this block matrix need not be square, hence the vector–Kronecker product multiplication algorithm in Sect. 2.2 involving rectangular factors.

We now turn to a class of problems that can be treated with approximative decompositional iterative methods using the understanding developed in this section.

4.3   Case Study from Closed Queueing Networks

We consider *closed* networks of first-come first-served (FCFS) queues with *phase-type* (PH) service distributions and arbitrary buffer sizes [56]. Queueing networks (QNs) have been used in the literature to model and analyze a variety of systems involving customers, packets, or jobs waiting to get service [90, 140]. Here, we concentrate on a relatively large class of problems that do not possess analytical solutions. Closedness implies that the number of customers circulating in the QN remains constant; there are no arrivals to the network from the outside, there are no departures to the outside, and the number of customers inside the network neither increases nor decreases as a result of the queueing discipline and the service process. A customer departs from a queue after getting service and joins a(nother) queue, possibly the same one it departed from. If a QN is not closed, it is said to be *open*. Regarding service distributions, *hypoexponential*, *hyperexponential*, *Coxian*, and *Erlang* are all PH and have rational Laplace transforms. Furthermore, the exponential distribution is a special case of the Erlang distribution, which is yet a special case of the hypoexponential distribution. Interestingly, it is proved that Erlang is the most suitable phase approximation for the deterministic distribution [2]. This is taken advantage of when modeling a robotic tape library [60] and a multiprocessor system [132] using SANs. For practical purposes, a five- to ten-phase Erlang is considered sufficient for approximating a deterministic distribution. The use of PH distributions in SANs is further investigated in [133]. The analysis of closed QNs with PH service distributions and arbitrary buffer sizes is challenging due to the fact that the corresponding state spaces grow exponentially with numbers of customers, queues, and phases in the service distribution of each queue. Detailed information regarding this work is available in [56, 108].

The network is defined by H queues, K customers, routing probability matrix P, PH service distribution $(T^{(h)}, \alpha^{(h)})$, where $T^{(h)}$ is the PH service distribution matrix of order t_h and $\alpha^{(h)}$ is the initial service probability distribution row vector of length t_h associated with $T^{(h)}$, and buffer size b_h for queue $h = 1, \ldots, H$. We start queue indices at one since we later use zero as the index of a special queue. This implies row and column indices of P also start from one. We let $c_h = \min\{K, b_h\}$ and represent the state of queue h by the ordered pair $i_h = (n_h, \phi_h)$, where $n_h = 0, \ldots, c_h$ denotes the occupancy of queue h and $\phi_h = 0, \ldots, t_h - 1$ denotes the phase of its service process, with the constraint that $\phi_h = 0$ when $n_h = 0$ (i.e., the phase is zero when the queue is empty). Then $i_h \in \{(0, 0)\} \bigcup \{1, \ldots, c_h\} \times \{0, \ldots, t_h - 1\}$. We remark that in our model, an arrival at a destination queue can only take place when the destination queue has space for the arriving customer; otherwise, the transition is inhibited. The implication of this assumption is that a customer will remain in the server until space becomes available in the destination queue.

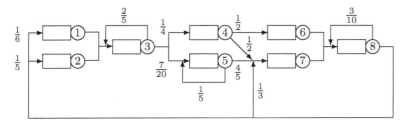

Fig. 4.1 Closed network of eight queues with phase-type service distributions and finite buffers

Example 3. Consider the closed QN with routing probability matrix

$$
P =
\begin{array}{c@{}c}
 & \begin{array}{cccccccc} 1 & 2 & 3 & 4 & 5 & 6 & 7 & 8 \end{array} \\
\begin{array}{c} 1 \\ 2 \\ 3 \\ 4 \\ 5 \\ 6 \\ 7 \\ 8 \end{array} &
\left(\begin{array}{cccccccc}
 & & 1 & & & & & \\
 & & 1 & & & & & \\
 & & & \frac{2}{5} & \frac{1}{4} & \frac{7}{20} & & \\
 & & & & & \frac{1}{2} & \frac{1}{2} & \\
 & & & & & \frac{1}{5} & \frac{4}{5} & \\
 & & & & & & & 1 \\
 & & & & & & & 1 \\
\frac{1}{6} & \frac{1}{5} & & & & & \frac{1}{3} & \frac{3}{10}
\end{array}\right)
\end{array}
$$

in Fig. 4.1 with eight queues (i.e., $H = 8$). The service distributions of the queues are given by

$$
T^{(1)}_{\text{Hypo}} = \begin{pmatrix} -\mu_0^{(1)} & \mu_0^{(1)} \\ & -\mu_1^{(1)} \end{pmatrix}, \qquad\qquad \alpha^{(1)} = e_1^T,
$$

$$
T^{(2)}_{\text{Hyper}} = \begin{pmatrix} -\mu_0^{(2)} & \\ & -\mu_1^{(2)} \end{pmatrix}, \qquad\qquad \alpha^{(2)} = \left(\alpha^{(2)}(0), \alpha^{(2)}(1) \right),
$$

$$
T^{(3)}_{\text{Erlang}} = \begin{pmatrix} -\zeta^{(3)} & \zeta^{(3)} & & & \\ & -\zeta^{(3)} & \zeta^{(3)} & & \\ & & -\zeta^{(3)} & \zeta^{(3)} & \\ & & & -\zeta^{(3)} & \zeta^{(3)} \\ & & & & -\zeta^{(3)} \end{pmatrix}, \qquad \alpha^{(3)} = e_1^T,
$$

$$
T^{(4)}_{\text{Hyper}} = \begin{pmatrix} -\mu_0^{(4)} & \\ & -\mu_1^{(4)} \end{pmatrix}, \qquad\qquad \alpha^{(4)} = \left(\alpha^{(4)}(0), \alpha^{(4)}(1) \right),
$$

$$
T^{(5)}_{\text{Hypo}} = \begin{pmatrix} -\mu_0^{(5)} & \mu_0^{(5)} \\ & -\mu_1^{(5)} \end{pmatrix}, \qquad\qquad \alpha^{(5)} = e_1^T,
$$

$$
T^{(6)}_{\text{Hypo}} = \begin{pmatrix} -\mu_0^{(6)} & \mu_0^{(6)} \\ & -\mu_1^{(6)} \end{pmatrix}, \qquad\qquad \alpha^{(6)} = e_1^T,
$$

$$T_{\text{Hyper}}^{(7)} = \begin{pmatrix} -\mu_0^{(7)} & \\ & -\mu_1^{(7)} \end{pmatrix}, \qquad\qquad \boldsymbol{\alpha}^{(7)} = \left(\boldsymbol{\alpha}^{(7)}(0), \boldsymbol{\alpha}^{(7)}(1) \right),$$

$$T_{\text{Erlang}}^{(8)} = \begin{pmatrix} -\zeta^{(8)} & \zeta^{(8)} & & & \\ & -\zeta^{(8)} & \zeta^{(8)} & & \\ & & -\zeta^{(8)} & \zeta^{(8)} & \\ & & & -\zeta^{(8)} & \zeta^{(8)} \\ & & & & -\zeta^{(8)} \end{pmatrix}, \quad \boldsymbol{\alpha}^{(8)} = \mathbf{e}_1^T.$$

Queues 1, 5, and 6 have two-phase hypoexponential service distributions, while queues 2, 4, and 7 have two-phase hyperexponential service distributions, and queues 3 and 8 have five-phase Erlang service distributions. In the two-phase hypoexponential service distribution, the customer takes an exponentially distributed service in phase 0 after which it takes a second exponentially distributed service in phase 1, and the parameters of the two exponential service distributions need not be the same. If the parameters of the two exponential distributions are the same, then the hypoexponential service distribution specializes to the two-phase Erlang service distribution. Since both processes start in phase 0, the initial probability distribution is 1 in phase 0 and 0 in others; hence, we have $\boldsymbol{\alpha}^{(h)} = \mathbf{e}_1^T$ for queue h with hypoexponential or Erlang service distribution. On the other hand, the hyperexponential distribution is a mix of exponential distributions specified by the initial probability distribution. That is, with probability $\boldsymbol{\alpha}^{(h)}(\phi_h)$ queue h takes an exponentially distributed service with rate $\mu_{\phi_h}^{(h)}$ such that $\sum_{\phi_h=0}^{t_h-1} \boldsymbol{\alpha}^{(h)}(\phi_h) = 1$.

The irreducible MC representing the evolution of queue $h = 1, \ldots, H$ is the $(c_h + 1) \times (c_h + 1)$ block tridiagonal matrix [19]

$$Q^{(h)} = G^{(h)} + D^{(h)},$$

where

$$G^{(h)} = \begin{pmatrix} G^{(h)}(0,0) & G^{(h)}(0,1) & & & \\ G^{(h)}(1,0) & G^{(h)}(1,1) & G^{(h)}(1,2) & & \\ & \ddots & \ddots & \ddots & \\ & & G^{(h)}(c_h-1, c_h-2) & G^{(h)}(c_h-1, c_h-1) & G^{(h)}(c_h-1, c_h) \\ & & & G^{(h)}(c_h, c_h-1) & G^{(h)}(c_h, c_h) \end{pmatrix},$$

$$\tag{4.1}$$

$G^{(h)}(0,0) = 0$, $\lambda^{(h)}(n_h)$ is the rate of arrivals to queue h under buffer occupancy n_h, $G^{(h)}(0,1) = \lambda^{(h)}(0)\boldsymbol{\alpha}^{(h)}$, $\mathbf{t}^{(h)} = -T^{(h)}\mathbf{e}$, $G^{(h)}(1,0) = \mathbf{t}^{(h)}$, $G^{(h)}(n_h, n_h) = T^{(h)}$ for $n_h = 1, \ldots, c_h$, $G^{(h)}(n_h, n_h - 1) = \mathbf{t}^{(h)}\boldsymbol{\alpha}^{(h)}$ for $n_h = 2, \ldots, c_h$, $G^{(h)}(n_h, n_h + 1) = \lambda^{(h)}(n_h)I_{t_h}$ for $n_h = 1, \ldots, c_h - 1$, and $D^{(h)}$ is the diagonal correction matrix summing the rows of $Q^{(h)}$ to zero. Superdiagonal blocks $G^{(h)}(n_h, n_h + 1)$ and subdiagonal blocks $G^{(h)}(n_h, n_h - 1)$ of $G^{(h)}$ represent service completions and arrivals of customers, respectively. Diagonal blocks $G^{(h)}(n_h, n_h)$ represent phase changes. The boundary level corresponding to a buffer occupancy of 0 has a single state, while the other levels each have t_h states. Hence, $G^{(h)}$ is a $(t_h c_h + 1) \times (t_h c_h + 1)$ matrix, and as we shall see in the next chapter, it is the infinitesimal generator of a finite LDQBD process.

Assuming that states of the irreducible MC underlying the closed QN are represented as $\mathbf{i} = (i_1, \ldots, i_H)$, let us define the mapping $f : \mathcal{S} \to \mathcal{N}$ as

$$f(\mathbf{i}) = f((i_1, \ldots, i_H)) = f(((n_1, \phi_1), \ldots, (n_H, \phi_H))) = (n_1, \ldots, n_H) = \mathbf{n}$$

for $\mathbf{i} \in \mathcal{S}$ and $\mathbf{n} = (n_1, \ldots, n_H) \in \mathcal{N}$. This mapping is onto and partitions $\mathcal{S}$ into equivalence classes. We denote the set of equivalence classes defined by f as $\mathcal{N}$, let $N = |\mathcal{N}|$, and remark that $\mathbf{n} = (n_1, \ldots, n_H) \in \mathcal{N}$ satisfies $\sum_{h=1}^{H} n_h = K$.

When queues modeled as in (4.1) are interconnected to form a closed QN, the arrival rate $\lambda^{(h)}(n_h)$ depends on column h of P, the states of the queues corresponding to nonzero elements in that column, and the rates by which the queues complete the last phases of their service processes for $h = 1, \ldots, H$. Hence, the value of $\lambda^{(h)}(n_h)$ for queue h is not known a priori but is to be determined as a result of steady-state analysis. Assuming the lexicographical order is associated with the states in $\mathcal{N}$, the generator matrix Q of such a closed QN can be expressed as an $(N \times N)$ block matrix with block $(\mathbf{n}, \mathbf{m})$ for $\mathbf{n}, \mathbf{m} \in \mathcal{N}$ given by [19]

$$Q(\mathbf{n}, \mathbf{m}) = \begin{cases} Q^{(h,l)}(\mathbf{n}, \mathbf{m}) & \text{if } \mathbf{n} \neq \mathbf{m} \text{ and } \mathbf{m} = \mathbf{n} - \mathbf{e}_h^T + \mathbf{e}_l^T, \\ D(\mathbf{n}, \mathbf{n}) + Q_D(\mathbf{n}, \mathbf{n}) + \sum_{h=1}^{H} Q^{(h,h)}(\mathbf{n}, \mathbf{n}) & \\ \quad \text{if } \mathbf{n} = \mathbf{m}, \\ 0 & \text{otherwise}, \end{cases} \quad (4.2)$$

where $h, l = 1, \ldots, H$, $\mathbf{m} = \mathbf{n} - \mathbf{e}_h^T + \mathbf{e}_l^T$ refers to service completion at queue h and arrival at queue l when in state $\mathbf{n}$ so as to make a transition to state $\mathbf{m}$, $D(\mathbf{n}, \mathbf{n})$ is the diagonal matrix summing block $\mathbf{n}$ of rows in Q to zero:

$Q^{(h,l)}(\mathbf{n}, \mathbf{m})$

$$= \begin{cases} p(h,l) \left(I_{c_{\mathbf{n},h}^{(h,l)}} \otimes G^{(h)}(n_h, n_h - 1) \otimes I_{r_{\mathbf{n},h}^{(h,l)}} \right) \left(I_{c_{\mathbf{n},l}^{(h,l)}} \otimes I_{t_l} \otimes I_{r_{\mathbf{n},l}^{(h,l)}} \right) & \text{if } h < l, \\ p(h,l) \left(I_{c_{\mathbf{n},l}^{(h,l)}} \otimes I_{t_l} \otimes I_{r_{\mathbf{n},l}^{(h,l)}} \right) \left(I_{c_{\mathbf{n},h}^{(h,l)}} \otimes G^{(h)}(n_h, n_h - 1) \otimes I_{r_{\mathbf{n},h}^{(h,l)}} \right) & \text{if } h > l, \\ p(h,h) \left(I_{c_{\mathbf{n},h}^{(h,h)}} \otimes G^{(h)}(n_h, n_h - 1) \right) \left(I_{t_h} \otimes I_{r_{\mathbf{n},h}^{(h,h)}} \right) & \text{if } h = l, \end{cases} \quad (4.3)$$

$$Q_D(\mathbf{n}, \mathbf{n}) = \sum_{h=1}^{H} I_{c_{\mathbf{n},h}^{(h,h)}} \otimes G^{(h)}(n_h, n_h) \otimes I_{r_{\mathbf{n},h}^{(h,h)}}, \quad (4.4)$$

$c_{\mathbf{n},j}^{(h,l)} = \prod_{u=1}^{j-1} \text{size}_{\mathbf{n},u}^{(h,l)}$ and $r_{\mathbf{n},j}^{(h,l)} = \prod_{u=j+1}^{H} \text{size}_{\mathbf{n},u}^{(h,l)}$ represent respectively products of column and row dimensions of matrices [see (2.4)], and

$$
\text{size}_{\mathbf{n},u}^{(h,l)} = \begin{cases}
\#\_\text{of\_rows}(G^{(u)}(n_u, n_u)) & \text{if } u \neq h \text{ and } u \neq l, \\
\#\_\text{of\_rows}(G^{(l)}(n_l, n_l + 1)) & \text{if } u = l \text{ and } l > h, \\
\#\_\text{of\_cols}(G^{(l)}(n_l, n_l + 1)) & \text{if } u = l \text{ and } l < h, \\
\#\_\text{of\_rows}(G^{(h)}(n_h, n_h - 1)) & \text{if } u = h \text{ and } h > l, \\
\#\_\text{of\_cols}(G^{(h)}(n_h, n_h - 1)) & \text{if } u = h \text{ and } h < l.
\end{cases} \qquad (4.5)
$$

Let us make some observations regarding (4.2). When $\mathbf{n} \neq \mathbf{m}$, $\mathbf{m} = \mathbf{n} - \mathbf{e}_h^T + \mathbf{e}_l^T$, and $h \neq l$ [see (4.3)], block $Q(\mathbf{n}, \mathbf{m})$ is nonzero if customers who depart from queue h can join queue l. This is the case if $p(h, l) \neq 0$ and $n_l < c_l$. Those blocks of $Q(\mathbf{n}, \mathbf{m})$ for which $p(h, l) = 0$ are zero. On the other hand, block $Q(\mathbf{n}, \mathbf{n})$ corresponds to the sum of the cases in which queues make phase transitions [second term, see (4.4)] and the departing customers from queues arrive at the same queues [third term, see (4.3)], plus a diagonal correction (first term) so as to sum up the rows to zero. For nonzero $Q^{(h,l)}(\mathbf{n}, \mathbf{m})$ in (4.3), including the case $h = l$, which implies $\mathbf{n} = \mathbf{m}$, there are three possibilities. First, the departure of a customer from a queue is an arrival at a queue with a larger index ($h < l$). Second, the departure of a customer from a queue is an arrival at a queue with a smaller index ($h > l$). Third, the departure of a customer from a queue is an arrival at the same queue ($h = l$).

Example 3 (cntd.). Let us consider the case of K customers for our closed QN with $H = 8$ queues and buffer sizes given by $\mathbf{b} = (8, 9, 9, 6, 6, 9, 9, 7)$. For up to six customers (i.e., $K \leq 6$), we obtain the maximum occupancies given by $\mathbf{c} = (K, K, K, K, K, K, K, K)$ since $c_h = \min\{K, b_h\}$ and $b_h \geq 6$ for $h = 1, \ldots, H$. This implies

$$
N = \binom{H + K - 1}{K} ;
$$

that is, Q has $(H + K - 1)$ choose K row and column of blocks when $\mathbf{c} = K\mathbf{e}^T$. For instance, when $K = 2$, we obtain $N = 36$. These 36 states in lexicographical order are $(0, 0, 0, 0, 0, 0, 0, 2)$, $(0, 0, 0, 0, 0, 0, 1, 1)$, $(0, 0, 0, 0, 0, 0, 2, 0)$, $(0, 0, 0, 0, 0, 1, 0, 1)$, $(0, 0, 0, 0, 0, 1, 1, 0)$, $(0, 0, 0, 0, 0, 2, 0, 0)$, $(0, 0, 0, 0, 1, 0, 0, 1)$, $\ldots$, $(2, 0, 0, 0, 0, 0, 0, 0)$. To each of these 36 states (or equivalence classes) there correspond multiple states from $\mathcal{S}$. For instance, the five states $((\underline{0}, 0), (\underline{0}, 0), (\underline{0}, 0), (\underline{0}, 0), (\underline{0}, 0), (\underline{0}, 0), (\underline{0}, 0), (\underline{2}, \phi_8)) \in \mathcal{S}$ for $\phi_8 = 0, \ldots, 4$ correspond to state $(0, 0, 0, 0, 0, 0, 0, 2) \in \mathcal{N}$ since queue 8 has an Erlang service distribution with five phases. On the other hand, the ten states $((\underline{0}, 0), (\underline{0}, 0), (\underline{0}, 0), (\underline{0}, 0), (\underline{0}, 0), (\underline{0}, 0), (\underline{1}, \phi_7), (\underline{1}, \phi_8)) \in \mathcal{S}$ for $\phi_7 = 0, 1$ and $\phi_8 = 0, \ldots, 4$ correspond to state $(0, 0, 0, 0, 0, 0, 1, 1) \in \mathcal{N}$ since queue 7 has a hyperexponential service distribution with two phases and queue 8 is as mentioned previously. Once K exceeds 6, N will be less than $(H + K - 1)$ choose K.

We now illustrate how the diagonal and off-diagonal blocks of Q respectively corresponding to $((0, 0, 0, 0, 0, 0, 0, 2), (0, 0, 0, 0, 0, 0, 0, 2))$ and $((0,0,0,0,0,0,1,1),(0,0,0,0,0,0,0,2))$ are obtained from (4.2)–(4.5) using the values in the routing probability matrix, P.

For the diagonal block we have

$$Q((0,0,0,0,0,0,0,2),(0,0,0,0,0,0,0,2))$$

$$= D((0,0,0,0,0,0,0,2),(0,0,0,0,0,0,0,2))$$

$$+Q_D((0,0,0,0,0,0,0,2),(0,0,0,0,0,0,0,2))$$

$$+\sum_{h=1}^{8} Q^{(h,h)}((0,0,0,0,0,0,0,2),(0,0,0,0,0,0,0,2))$$

$$= D((0,0,0,0,0,0,0,2),(0,0,0,0,0,0,0,2))$$

$$+\sum_{h=1}^{8} I_{c_{(0,0,0,0,0,0,0,2),h}^{(h,h)}} \otimes G^{(h)}(n_h,n_h) \otimes I_{r_{(0,0,0,0,0,0,0,2),h}^{(h,h)}}$$

$$+\sum_{h=1}^{8} p(h,h)\left(I_{c_{(0,0,0,0,0,0,0,2),h}^{(h,h)}} \otimes G^{(8)}(n_h,n_h-1)\right)\left(I_{t_h} \otimes I_{r_{(0,0,0,0,0,0,0,2),h}^{(h,h)}}\right)$$

$$= D((0,0,0,0,0,0,0,2),(0,0,0,0,0,0,0,2))$$

$$+I_{c_{(0,0,0,0,0,0,0,2),8}^{(8,8)}} \otimes G^{(8)}(2,2) \otimes I_{r_{(0,0,0,0,0,0,0,2),8}^{(8,8)}}$$

$$+p(8,8)\left(I_{c_{(0,0,0,0,0,0,0,2),8}^{(8,8)}} \otimes G^{(8)}(2,1)\right)\left(I_{t_8} \otimes I_{r_{(0,0,0,0,0,0,0,2),8}^{(8,8)}}\right)$$

$$= D((0,0,0,0,0,0,0,2),(0,0,0,0,0,0,0,2))$$

$$+I_1 \otimes G^{(8)}(2,2) \otimes I_1 + p(8,8)(I_1 \otimes G^{(8)}(2,1))(I_{t_8} \otimes I_1)$$

$$= D((0,0,0,0,0,0,0,2),(0,0,0,0,0,0,0,2)) + T^{(8)}_{\text{Erlang}} + \frac{3}{10}\mathbf{t}^{(8)}\boldsymbol{\alpha}^{(8)} I_5$$

$$= D((0,0,0,0,0,0,0,2),(0,0,0,0,0,0,0,2)) + T^{(8)}_{\text{Erlang}} + \frac{3}{10}\boldsymbol{\zeta}^{(8)}\mathbf{e}_8\mathbf{e}_1^T$$

$$= \begin{pmatrix} * & \zeta^{(8)} & & & \\ & * & \zeta^{(8)} & & \\ & & * & \zeta^{(8)} & \\ & & & * & \zeta^{(8)} \\ \frac{3}{10}\zeta^{(8)} & & & & * \end{pmatrix}.$$

Observe that the first summation has a single term that contributes to the result since $G^{(h)}(n_h,n_h) = G^{(h)}(0,0) = 0$ for $h = 1,\ldots,7$. In other words, no phase change in the service process is possible without an arrival if the queue is empty. Similarly, the second summation has a single term that contributes to the result since $n_h = 0$

for $h = 1, \ldots, 7$. That is, a departure from a queue joins the same queue only if there is a customer taking service in the last phase, the service ends, the customer departs, and there is a positive probability of joining the same queue.

For the off-diagonal block we have

$$Q((0,0,0,0,0,0,1,1),(0,0,0,0,0,0,0,2))$$

$$= Q^{(7,8)}((0,0,0,0,0,0,1,1),(0,0,0,0,0,0,0,2))$$

$$= p(7,8)\left(I_{c^{(7,8)}_{(0,0,0,0,0,0,1,1),7}} \otimes G^{(7)}(1,0) \otimes I_{r^{(7,8)}_{(0,0,0,0,0,0,1,1),7}}\right)$$

$$\left(I_{c^{(7,8)}_{(0,0,0,0,0,0,1,1),8}} \otimes I_{t_8} \otimes I_{r^{(7,8)}_{(0,0,0,0,0,0,1,1),8}}\right)$$

$$= (I_1 \otimes G^{(7)}(1,0) \otimes I_{t_8})(I_{t_7} \otimes I_{t_8} \otimes I_1)$$

$$= (\mathbf{t}^{(7)} \otimes I_5)(I_2 \otimes I_5) = \begin{pmatrix} \mu_0^{(7)} \\ \mu_1^{(7)} \end{pmatrix} \otimes I_5 = \begin{pmatrix} \mu_0^{(7)} & & & & \\ & \mu_0^{(7)} & & & \\ & & \mu_0^{(7)} & & \\ & & & \mu_0^{(7)} & \\ & & & & \mu_0^{(7)} \\ \mu_1^{(7)} & & & & \\ & \mu_1^{(7)} & & & \\ & & \mu_1^{(7)} & & \\ & & & \mu_1^{(7)} & \\ & & & & \mu_1^{(7)} \end{pmatrix}.$$

In practice, Q is neither generated nor stored; instead, the efficient vector–Kronecker product multiplication algorithm in Sect. 2.2 is used to carry out the steady-state analysis of Q underlying the closed QN. To that end, any method discussed in Chap. 3 can be used. Here we address two approximative decompositional methods, the first of which appears in [107] and the second one in [152]. We show how these two iterative methods for closed QNs can be implemented using Kronecker products. Together with [56], [108] provides detailed information regarding this work.

In its setup phase, the first method partitions the closed QN into subnetworks. In doing this, it classifies queue h for $h = 1, \ldots, H$ as finite or infinite buffer depending on whether or not $b_h < K$. Finite-buffer queues are those that have positive blocking probabilities for given K. The method places queues feeding a finite-buffer queue in the same subnetwork as the finite-buffer queue. In this way, the method aims at achieving a decomposition in which transition probabilities between subnetworks are independent of the states of the subnetworks. Thus, each subnetwork can be considered as a service station with a state-dependent exponential service rate for which the parameters of the equivalent server are obtained by analyzing the subnetwork in isolation as an open QN assuming it has state-dependent Poisson arrivals. The approximate results are obtained via fixed-point iteration, which requires throughputs of subnetworks to be computed.

Now let us assume that the set of queue indices $\mathcal{H} = \{1, \ldots, H\}$ of the closed QN is partitioned into J subsets, with $\mathcal{H}^{(j)}$ corresponding to subnetwork j for $j = 1, \ldots, J$. Hence, $\mathcal{H} = \bigcup_{j=1}^{J} \mathcal{H}^{(j)}$ and $\mathcal{H}_j \cap \mathcal{H}_l = \emptyset$ for $j \neq l$, $j, l = 1, \ldots, J$.

Example 3 (cntd.). The closed QN in Fig. 4.1 with $\mathbf{b} = (8, 9, 9, 6, 6, 9, 9, 7)$ has eight subnetworks of single queues in $\{\{1\}, \{2\}, \{3\}, \{4\}, \{5\}, \{6\}, \{7\}, \{8\}\}$ for $K = 6$, five subnetworks of single queues and one subnetwork of three queues in $\{\{1\}, \{2\}, \{3, 4, 5\}, \{6\}, \{7\}, \{8\}\}$ for $K = 7$, two subnetworks of single queues and two subnetworks of three queues in $\{\{1\}, \{2\}, \{3, 4, 5\}, \{6, 7, 8\}\}$ for $K = 8$, and one subnetwork of two queues and two subnetworks of three queues in $\{\{1, 2\}, \{3, 4, 5\}, \{6, 7, 8\}\}$ for $K = 9$. Hence, we have $J = 8$, $J = 6$, $J = 4$, and $J = 3$ for $K = 6, 7, 8, 9$, respectively.

We denote by K_j the number of customers in subnetwork j defined by $\mathcal{H}^{(j)}$ and by

$$\mathbf{v}(j) = \sum_{h \in \mathcal{H}^{(j)}} \xi(h) \left(\sum_{l \in \mathcal{H} \setminus \mathcal{H}^{(j)}} p(h, l) \right) \quad \text{for } j = 1, \ldots, J$$

the visit ratio of subnetwork j for which ξ is the solution of $\xi P = \xi$ subject to $\xi(1) = 1$. Note that P is expected to be irreducible for closed QNs. Furthermore, we let $\psi_{(m)}^{(j)}(k)$, $\lambda_{(m)}^{(j)}(k)$, and $\pi_{(m)}^{(j)}(k)$ denote approximate throughput, Poisson arrival rate, and steady-state probability of the open QN defined by $\mathcal{H}^{(j)}$ with k customers at iteration m of the fixed-point iteration, respectively. The method starts by initializing the throughput values $\psi_{(0)}^{(j)}(k)$ of subnetwork j for $k = 1, \ldots, K$. For this purpose, we use an analytical method, the convolution algorithm (CA) [79].

At iteration m, for $j = 1, \ldots, J$ the first approximative decompositional method is defined by the system of equations [107]

$$\sigma_{(m)}^{(j)}(K_j) = \begin{cases} 1 & \text{if } K_j = 0, \\ \prod_{k=1}^{K_j} \psi_{(m)}^{(j)}(k) & \text{if } K_j > 0; \end{cases}$$

$$\beta_{(m)}^{(j)}(u) = \sum_{K_1, \ldots, K_J} \left(\prod_{j'=1}^{J} \frac{\mathbf{v}(j')^{K_{j'}}}{\sigma_{(m)}^{(j')}(K_{j'})} \right) \quad \text{and} \quad \sum_{j'=1}^{J} K_{j'} = u, \; K_j = 0;$$

$$\lambda_{(m)}^{(j)}(l) = \mathbf{v}(j) \frac{\beta_{(m)}^{(j)}(K - l - 1)}{\beta_{(m)}^{(j)}(K - l)} \quad \text{for } l = 0, \ldots, K - 1;$$

$$\psi_{(m+1)}^{(j)}(k) = \mathbf{v}(j) \frac{\beta_{(m)}^{(j)}(K - k)}{\beta_{(m)}^{(j)}(K - k + 1)} \frac{\pi_{(m)}^{(j)}(k - 1)}{\pi_{(m)}^{(j)}(k)} \quad \text{for } k = 1, \ldots, K,$$

Fig. 4.2 Open QN modeled
as closed QN with slack
queue having a buffer size
of K

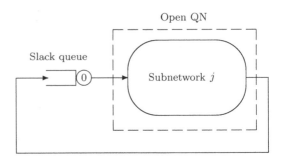

where $\beta_{(m)}^{(j)}(u)$ is the normalization constant associated with subnetwork j under
the condition that it has no customers and the other subnetworks have a total of
u customers. The arrival rate and steady-state vectors for $\mathcal{H}^{(j)}$ at iteration m are
given by $\boldsymbol{\lambda}_{(m)}^{(j)} = (\lambda_{(m)}^{(j)}(0),\ldots,\lambda_{(m)}^{(j)}(K-1))$ and $\boldsymbol{\pi}_{(m)}^{(j)} = (\pi_{(m)}^{(j)}(0),\ldots,\pi_{(m)}^{(j)}(K))$,
respectively.

At iteration m, subnetwork j is perceived as an open QN with the state-
dependent Poisson arrival rates in $\boldsymbol{\lambda}_{(m)}^{(j)}$ and analyzed for its steady-state vector.
To this end, the open QN is modeled as a closed QN, which consists of the
subnetwork's queues in $\mathcal{H}^{(j)}$ and a slack queue numbered 0 as in Fig. 4.2. The
slack queue is assumed to have a buffer size of K and can be considered infinite,
therefore nonblocking. It mimics the state-dependent Poisson arrivals of customers
at subnetwork j with the rates in $\boldsymbol{\lambda}_{(m)}^{(j)}$ and has an exponentially distributed
service with state-dependent service rates in $\boldsymbol{\psi}_{(m)}^{(j)}$. The state-dependent throughputs
are the state-dependent service rates since the slack queue is practically infinite.
Each closed QN obtained as such can be modeled by defining the queues in the
subnetwork and the slack queue following (4.1) and then constructing the block
matrix that represents the interactions among the queues in the closed QN using
Kronecker products as in (4.2)–(4.5). Consequently, one can employ the ML method
to compute $\boldsymbol{\pi}_{(m)}^{(j)}$ for $j = 1,\ldots,J$ and then obtain $\boldsymbol{\psi}_{(m+1)}^{(j)}$ to prepare for the next
iteration.

On the other hand, the second decompositional method partitions the closed
QN into individual queues and approximates the service distribution of each queue
by a state-dependent exponential service distribution. Thus, the method transforms
the closed QN into another closed QN with state-dependent exponential service
distributions. The decomposition in this approach is maximal, meaning each queue
is placed in a separate subnetwork (i.e., $J = H$).

Example 3. (cntd.) The closed QN in Fig. 4.1 is partitioned into eight subnetworks
defined by $\mathcal{H}^{(j)} = \{j\}$ for $j = 1,\ldots,8$.

As in Fig. 4.2, a slack queue with an infinite buffer and a state-dependent
exponential service distribution is used to model state-dependent Poisson arrivals at
queue j in $\mathcal{H}^{(j)}$ having a PH service distribution for $j = 1,\ldots,H$. Following this
approximation, the method sets the state-dependent service rate of the slack queue

to some initial value and then employs a fixed-point iteration on the decomposed network to compute the throughputs of all queues. Again, initialization of the state-dependent service rates of slack queues, which are their state-dependent throughputs, can be done using CA.

At iteration m, for $j = 1, \ldots, H$ the fixed-point iteration of the second approximative decompositional method is based on the system of equations [152]

$$\bar{\pi}_{(m)} \bar{Q}_{(m)} = \mathbf{0} \quad \text{subject to} \quad \bar{\pi}_{(m)} \mathbf{e} = 1,$$

$$\lambda_{(m)}^{(j)}(l) = \psi_{(m)}^{(j)}(l+1) \frac{\bar{\pi}_{(m)}^{(j)}(l+1)}{\bar{\pi}_{(m)}^{(j)}(l)} \quad \text{for } l = 0, \ldots, c_j - 1,$$

$$\psi_{(m+1)}^{(j)}(k) = \lambda_{(m)}^{(j)}(k-1) \frac{\pi_{(m)}^{(j)}(k-1)}{\pi_{(m)}^{(j)}(k)} \quad \text{for } k = 1, \ldots, c_j,$$

where $\psi_{(m)}^{(j)}(k)$ and $\lambda_{(m)}^{(j)}(k)$ denote approximate throughput and Poisson arrival rate of the open QN defined by $\mathcal{H}^{(j)}$ with k customers, respectively. Furthermore, $\bar{\pi}_{(m)}$ is the steady-state vector of the infinitesimal generator $\bar{Q}_{(m)}$ underlying the closed QN constructed by replacing the PH service distribution of queue j in the original closed QN with state-dependent exponential service distribution having rate $\psi_{(m)}^{(j)}(k)$ for $j = 1, \ldots, H$. Here, $\bar{\pi}_{(m)}^{(j)}(l)$ is the marginal steady-state probability of having l customers in queue j in the closed QN represented by $\bar{Q}_{(m)}$. Finally, $\pi_{(m)}^{(j)}$ is the steady-state vector of subnetwork j and $\pi_{(m)}^{(j)}(k)$ is the marginal steady-state probability of having k customers in subnetwork j. Hence, each closed QN obtained as such can be modeled by defining the queue in the subnetwork and the slack queue following (4.1) and then constructing the block matrix that represents the interactions among the two queues in the closed QN using Kronecker products as in (4.2)–(4.5). Consequently, one can employ the ML method to compute $\pi_{(m)}^{(j)}$ for $j = 1, \ldots, J$ and then obtain $\psi_{(m+1)}^{(j)}$ to prepare for the next iteration.

Implementations of the two approximative decompositional methods (respectively called M and YB henceforth) are available in Matlab [109] together with implementations of CA, a mean value analysis algorithm for blocking closed QNs (MVABLO) [1], point iterative methods based on splittings, and ML with fixed and circular orders of aggregation in V-, F-, and W-cycles. Experiments are performed on a 3-GHz Pentium processor with 1 GB of memory. The methods are compared for their accuracy and efficiency on various models for analyzing utilizations and mean queue lengths. ML and GS methods assume stopping tolerances of 10^{-15} on the residual 1-norm. ML uses GS as the smoother and performs one pre- and one postsmoothing at each level. A stopping tolerance of 10^{-4} is used on the approximate error of utilizations and mean lengths of queues for M and YB. The subnetworks resulting from decomposition in these methods are solved with ML. When computing the steady-state vector of the coarsest infinitesimal generator in

M and the steady-state vector of the state-dependent closed QN with exponential service distributions in YB, if the order of the matrix is less than 500, Gaussian elimination (GE) otherwise BICGSTAB with ILU preconditioning and a drop tolerance of 10^{-5} [59], is used. The results obtained by approximative methods are compared with the results of ML, and relative errors are provided using the 1-norm. Note that relative errors are indicative of numbers of correct decimal digits in the results. That is, an approximate result with a relative error on the order of 10^{-z} implies z correct decimal digits.

Example 3 (ctnd.). The closed QN of Fig. 4.1 is considered with two sets of real-valued parameters for its PH service distributions. Since buffer sizes of queues are finite, we may have different subnetwork topologies in M for different numbers of customers. In this regard, balanced and unbalanced service requirements in the subnetworks are considered.

For *balanced* service requirements we choose

$$\left(\mu_0^{(1)}, \mu_1^{(1)}\right) = (1.5, 2), \qquad \left(\mu_0^{(2)}, \mu_1^{(2)}\right) = (1.1, 1.1), \quad \boldsymbol{\alpha}^{(2)} = (1/10, 9/10),$$

$$\zeta^{(3)} = 18, \qquad \left(\mu_0^{(4)}, \mu_1^{(4)}\right) = (0.9, 0.9), \quad \boldsymbol{\alpha}^{(4)} = (17/20, 3/20),$$

$$\left(\mu_0^{(5)}, \mu_1^{(5)}\right) = (1.5, 10), \quad \left(\mu_0^{(6)}, \mu_1^{(6)}\right) = (0.5, 4), \quad \left(\mu_0^{(7)}, \mu_1^{(7)}\right) = (3.5, 3.5),$$

$$\boldsymbol{\alpha}^{(7)} = (1/4, 3/4), \quad \zeta^{(8)} = 30.$$

For *unbalanced* service requirements we choose

$$\left(\mu_0^{(1)}, \mu_1^{(1)}\right) = (200, 90), \qquad \left(\mu_0^{(2)}, \mu_1^{(2)}\right) = (1000, 1000),$$

$$\boldsymbol{\alpha}^{(2)} = (1/10, 9/10), \quad \zeta^{(3)} = 0.15,$$

$$\left(\mu_0^{(4)}, \mu_1^{(4)}\right) = (0.008, 0.008), \quad \boldsymbol{\alpha}^{(4)} = (17/20, 3/20),$$

$$\left(\mu_0^{(5)}, \mu_1^{(5)}\right) = (8000, 5000), \quad \left(\mu_0^{(6)}, \mu_1^{(6)}\right) = (0.1, 0.05),$$

$$\left(\mu_0^{(7)}, \mu_1^{(7)}\right) = (10, 10), \qquad \boldsymbol{\alpha}^{(7)} = (1/4, 3/4),$$

$$\zeta^{(8)} = 30.$$

Note that the initial probability vectors of PH service distributions are identical for the balanced and unbalanced service requirements of subnetworks.

We first consider a closed QN with buffer sizes in $\mathbf{b} = (8, 9, 9, 6, 6, 9, 9, 7)$. The underlying CTMCs have state-space sizes of 85,991, 236,172, 578,592, 1,291,130 and infinitesimal generators with 1,716, 3,430, 6,418, 11,359 blocks in each dimension for $K = 6, 7, 8, 9$ customers, respectively. For all values of K in the balanced case, MVABLO yields 1.5 digits of accuracy for both performance measures. The results of CA are not better than those of MVABLO. Utilizations obtained with YB are 0.5 digits more accurate than those of M, yet YB executes more flops than M. Furthermore, YB executes more flops than ML for $K = 7, 9$; hence, YB is not useful in these cases. M provides 2 to 2.5 digits of accuracy in the results and becomes, if not the more accurate, the more efficient method when compared with YB. For none of the values of K considered does GS perform fewer flops than ML. As for the unbalanced case, CA and MVABLO yield 2 digits of accuracy for utilizations and 2.5 digits of accuracy for mean lengths of queues when $K = 6$. By performing a negligible number of flops, CA and MVABLO emerge as the most accurate and efficient methods in this case. However, as K increases, M yields at least 3.5 digits of accuracy for both performance measures, while CA and MVABLO are able to provide at most 1.5 digits of accuracy. YB preserves at least 2 digits of accuracy for utilizations and 2.5 digits of accuracy for mean lengths of queues when $K > 6$. Yet, the flop counts of M are less than those of YB, and thus M comes across as the more accurate and efficient method of the two. GS does not converge to the prespecified tolerance within a reasonable number of iterations or time.

Another set of experiments is performed with $\mathbf{b} = (5, 9, 9, 5, 5, 9, 9, 5)$. In this case, the closed QNs have state-space sizes of 85,980, 235,960, 576,670, 1,279,970 and infinitesimal generators with 1,712, 3,400, 6,291, 10,960 blocks in each dimension for $K = 6, 7, 8, 9$ customers, respectively. For all values of K in the balanced case, MVABLO yields 1.5 digits of accuracy for both performance measures. CA yields 1 digit of accuracy for mean queue lengths in all cases and 1 digit of accuracy for utilizations in all cases except $K = 7$. Utilizations obtained with M yield 2.5 digits of accuracy. Those obtained with YB are equally good or slightly better. On the other hand, mean queue lengths obtained with YB yield 2 digits of accuracy; those obtained with M are equally good or slightly better. Since ML converges within ten iterations in all cases, YB ends up performing more flops than ML for $K > 6$, while M performs 5 to 20 times fewer flops than ML. Additionally, when we consider the accuracy of M and compare its flop counts with those of ML and YB, we conclude that M is the most efficient method. Regarding the unbalanced case, CA and MVABLO yield 1.5 digits of accuracy for utilizations and 0.5–1.5 digits of accuracy for mean lengths of queues, respectively. Results with YB are 2 digits accurate in all cases. On the other hand, results with M provide 2.5 to 4 digits of accuracy for both performance measures, and M performs at least five times fewer flops than YB. Hence, M emerges again as the more accurate and efficient method of the two.

A total of 67 problems arising from 5 different closed QNs with PH service distributions and arbitrary buffer sizes (including the 16 problems associated with Example 3) are analyzed for utilizations and mean lengths of queues using the software tool. Three of the problems have 6 queues, whereas the other two have 3 and 8 queues, the latter corresponding to Fig. 4.1. CA and MVABLO produce acceptable results for problems with balanced service requirements and a relatively small number of blocking queues. On the other hand, M and YB provide relatively more accurate results for all problems and yield results with at least two digits of accuracy for unbalanced service requirements. Also, unlike the results obtained with CA and MVABLO, an increase in the number of blocking queues has almost no effect on the results obtained with M and YB. Therefore, M and YB emerge as more accurate methods than CA and MVABLO for problems with unbalanced service requirements and a relatively large number of blocking queues. When the accuracies of M and YB are compared, especially in problems with unbalanced service requirements and a relatively large number of blocking queues, we see that M can produce at least half a digit more accurate results for utilizations than YB.

When efficiencies of M and YB are compared, we see that the number of flops performed by YB to compute the arrival rates of queues mostly depends on the number of flops performed for obtaining the solution of the state-dependent closed QN with exponential service distributions generated at each fixed-point iteration. Therefore, for problems that require a relatively small number of flops for the solution of this QN, YB executes fewer flops than M. Also, for problems that result in subnetworks with a relatively large number of queues for M, YB may end up performing fewer flops than M through its fixed-point approximation process. Consequently, efficiencies of M and YB depend heavily on the particular problem. Nevertheless, the average number of fixed-point iterations performed by M and YB over all problems is four and five, respectively. When ML and GS are compared, we see that ML achieves convergence within 100 iterations in all problems. On the other hand, GS does not converge within a reasonable number of iterations or time in some of the problems. Clearly, the number of iterations determines the number of flops executed by the methods, and ML performs fewer flops than GS in almost all problems. Even though GS takes less space in memory than ML, in most of the problems ML requires less memory than the sparse representation of the infinitesimal generator underlying the closed QN with PH service distributions and arbitrary buffer sizes, which means it is capable of solving variants of problems with relatively large numbers of customers. Since M and YB are based on decomposition, the space requirements of M and YB are smaller than those of ML and GS for relatively large problems. Indeed, it is verified that the usage of ML in M and YB introduces another dimension of scalability to the space requirements of the two methods.

Chapter 5
Matrix-Analytic Methods

Matrix-analytic methods are geared toward MCs having state spaces that can be partitioned into subsets called *levels*. The transition matrices of such MCs, when symmetrically permuted according to increasing level number, should also have a particular nonzero structure, such as block tridiagonal or block Hessenberg. For instance, the well-known *quasi-birth-and-death* (QBD) processes fall into the class of processes that lend themselves to steady-state analysis with matrix analytic methods. These methods were originally proposed by Neuts [113, 114] for phase-type processes and have been improved over the years [14, 82, 101]. They characterize the solution by matrices having stochastic interpretations and sizes determined by the number of states within levels. Here, we consider CTMCs and concentrate on the class of *level-dependent* QBD (LDQBD) processes. We show how systems of stochastic chemical kinetics can be modeled using infinite LDQBD processes, expressed in the form of Kronecker products, and analyzed for their steady state with the help of Lyapunov theory. In passing, we remark that the concept of level introduced here has nothing to do with the level concept introduced during the discussion of iterative solution methods.

5.1 Level-Dependent Quasi-Birth-and-Death Processes

Continuous-time LDQBD processes are CTMCs having transition rate matrices that can be symmetrically permuted to the block tridiagonal form

$$Q = \begin{pmatrix} Q_{0,0} & Q_{0,1} & & & \\ Q_{1,0} & Q_{1,1} & Q_{1,2} & & \\ & \ddots & \ddots & \ddots & \\ & & Q_{l,l-1} & Q_{l,l} & Q_{l,l+1} \\ & & & \ddots & \ddots & \ddots \end{pmatrix}. \tag{5.1}$$

As opposed to QBD processes, the dependency on level number, l, in (5.1) manifests itself with two subscripts rather than one under each nonzero block for $l \in \mathbb{Z}_+$. It is the first subscript that corresponds to level number. Nevertheless, nonzero blocks can be dependent on level number in two different ways. It is either the nonzero values or the dimensions of a nonzero block (or both) that depend on level number. In this respect, LDQBD processes generalize QBD processes with nonhomogeneous transition rates and rectangular subdiagonal/superdiagonal nonzero blocks. Note that in this chapter we are using a slightly different notation in that we are indicating the blocks of Q with subscripts rather than in parentheses. As we shall see, this is so because the semantics of a block index here is different than what we have been using thus far, and this choice considerably simplifies the notation.

Assuming that the subset of states corresponding to level l is denoted by $\mathcal{S}_l$, the nonzero blocks at level l are given by

$$Q_{l,l-1} \in \mathbb{R}_{\geq 0}^{|\mathcal{S}_l| \times |\mathcal{S}_{l-1}|}, \, Q_{l,l} \in \mathbb{R}^{|\mathcal{S}_l| \times |\mathcal{S}_l|}, \, Q_{l,l+1} \in \mathbb{R}_{\geq 0}^{|\mathcal{S}_l| \times |\mathcal{S}_{l+1}|}.$$

Negative entries appear only along the diagonal of $Q_{l,l}$. There are a countably infinite number of levels, and transitions from level l are either to states within level l or to states in the adjacent levels $(l - 1)$ and $(l + 1)$. Level 0 is an exception since it constitutes the boundary level and has two nonzero blocks. Clearly, the ordering of states within a level is fixed only up to a permutation.

Assuming that steady state exists, its probability distribution vector may be written in a piecemeal manner as

$$\pi = (\pi^{(0)}, \pi^{(1)}, \ldots),$$

and its subvector at level $(l + 1)$ can be obtained from

$$\pi^{(l+1)} = \pi^{(l)} R_l \tag{5.2}$$

once the *matrix of conditional expected sojourn times* at level l

$$R_l = Q_{l,l+1}(-Q_{l+1,l+1} - R_{l+1}Q_{l+2,l+1})^{-1} \tag{5.3}$$

is available for $l \in \mathbb{Z}_+$ [18]. In (5.3), $R_l(\mathbf{i}, \mathbf{j})$ denotes the expected sojourn time in state $\mathbf{j} \in \mathcal{S}_{l+1}$ per unit sojourn in state $\mathbf{i} \in \mathcal{S}_l$ before returning to level l, given that the process started in state $\mathbf{i}$ [126]. We remark that

$$R_l \in \mathbb{R}_{\geq 0}^{|\mathcal{S}_l| \times |\mathcal{S}_{l+1}|} \quad \text{for} \quad l \in \mathbb{Z}_+$$

is nonnegative and rectangular. The recurrence in (5.2) requires $\pi^{(0)}$ to be determined first. This can be done from the set of boundary equations

$$\pi^{(0)}Q_{0,0} + \pi^{(1)}Q_{1,0} = 0$$

corresponding to the first column of blocks using $\boldsymbol{\pi}^{(1)} = \boldsymbol{\pi}^{(0)} R_0$ from (5.3). Hence, we conclude that $\boldsymbol{\pi}^{(0)}$ should be the positive left eigenvector of $(Q_{0,0} + R_0 Q_{1,0})$ corresponding to the eigenvalue 0 [111]. Eventually, $\boldsymbol{\pi}$ should be normalized so that $\boldsymbol{\pi} \mathbf{e} = 1$.

The next section discusses how the countable infiniteness of the state space $\mathcal{S} = \bigcup_{l \in \mathbb{Z}_+} \mathcal{S}_l$ can be handled during steady-state analysis when Q is irreducible. Note that we have not discussed how states are assigned to levels. We do that in a case study following the next section.

5.2 Handling Infiniteness

For multidimensional Markovian systems, it is most natural to model classes of transitions among states by using state-change vectors. In the next section, we provide a formal definition of transition classes that can be used in Kronecker-based modeling of systems of stochastic chemical kinetics. For the time being, we will be content with the more informal definition provided next.

Given a state vector $\mathbf{i} \in \mathbb{Z}_+^{1 \times H}$ representing the current state of an H-dimensional Markovian system, a *state-change vector* $\mathbf{v}^{(k)} \in \mathbb{Z}^{1 \times H}$ and a transition rate α_k : $\mathcal{S} \to \mathbb{R}_{\geq 0}$, the system makes a transition of class k to state $(\mathbf{i} + \mathbf{v}^{(k)})$ with rate $\alpha_k(\mathbf{i})$ when $\mathbf{i} \in \mathcal{S}$, $\alpha_k(\mathbf{i}) > 0$, and $\mathbf{i} + \mathbf{v}^{(k)} \geq \mathbf{0}$ for $k = 1, \ldots, K$. Here, we assume that $\mathbf{v}^{(k)} \neq \mathbf{0}$ and $\alpha_k(\mathbf{i})$ is a multivariate function in the state variables i_h for $h = 1, \ldots, H$ and $k = 1, \ldots, K$. It should be apparent from the values $\mathbf{i} \in \mathcal{S}$ can take that the state space $\mathcal{S}$ is countably infinite and an LDQBD model requires us to truncate it judiciously for analysis purposes. The following approach enables us to identify those states on which a certain amount of the steady-state probability mass is concentrated when $\mathcal{S}$ is irreducible.

An irreducible CTMC with countably infinite state space $\mathcal{S}$ has a steady-state distribution if and only if there exists a *Lyapunov function* $g : \mathcal{S} \to \mathbb{R}_{\geq 0}$ and a finite set $\mathcal{C} \subset \mathcal{S}$ simultaneously satisfying the three conditions [145]:

(1) $d(\mathbf{i}) \leq -\gamma$ for all $\mathbf{i} \in \mathcal{S} \setminus \mathcal{C}$ and some $\gamma > 0$,

(2) $d(\mathbf{i}) < \infty$ for all $\mathbf{i} \in \mathcal{C}$,

(3) $\{\mathbf{i} \in \mathcal{S} \mid g(\mathbf{i}) \leq r\}$ is finite for all $r < \infty$,

where

$$d(\mathbf{i}) = \sum_{k=1}^{K} \alpha_k(\mathbf{i})(g(\mathbf{i} + \mathbf{v}^{(k)}) - g(\mathbf{i})) \in \mathbb{R} \tag{5.4}$$

is the *drift* in state $\mathbf{i} \in \mathcal{S}$. In other words, the drift must be negative for the countably infinite number of states in the state space $\mathcal{S}$ excluding those in $\mathcal{C}$, the drift must be finite for the states in $\mathcal{C}$, and the set of states for which the Lyapunov function attains a finite value must always be finite.

When there is a candidate Lyapunov function $g(\mathbf{i})$ satisfying condition (3) at hand and $c = \sup_{\mathbf{i} \in \mathcal{S}} d(\mathbf{i}) < \infty$ [the latter condition is equivalent to satisfying conditions (1) and (2) simultaneously], it is possible to specify $\varepsilon = c/(c + \gamma) \in (0, 1)$ as an upper bound on $\sum_{\mathbf{i} \in \mathcal{S} \setminus \mathcal{C}} \pi(\mathbf{i})$. Equivalently, it is possible to use $\gamma = c/\varepsilon - c$ in constructively defining $\mathcal{C} = \{\mathbf{i} \in \mathcal{S} \mid -\gamma < d(\mathbf{i}) < \infty\}$. Now, if it is further shown that $\mathcal{C}$ is finite, then the three conditions given above hold and $\sum_{\mathbf{i} \in \mathcal{C}} \pi(\mathbf{i}) \geq 1 - \varepsilon$. However, in many cases it is not a trivial task to show the finiteness of $\mathcal{C}$ for a chosen Lyapunov function $g(\mathbf{i})$. This seems to be an open problem and worth exploring.

Clearly, determining c is also a problem in itself. To that end, the domain of the search for extrema should be restricted to $\mathbb{R}_{\geq 0}^{1 \times H}$. All extrema can then be computed by equating the gradient of $d(\mathbf{i})$ to zero. To determine all local extrema including those located on the boundaries of the domain, the same system must be solved for every projection of $d(\mathbf{i})$ onto each subspace of $\mathbb{R}^{1 \times H}$ by setting all combinations of state variables i_h for $h = 1, \ldots, H$ to 0. In the end, all extrema outside $\mathbb{R}_{\geq 0}^{1 \times H}$ should be discarded. Throughout this process, the resulting nonlinear equation systems can be solved, for instance, using the HOM4PS-2.0 package [102], which implements the polyhedral homotopy continuation method.

Once we have proved the finiteness of $\mathcal{C}$ and determined c (or, equivalently, γ for chosen ε), we can compute the pair of level numbers, $(Low, High)$, of the LDQBD process such that the states within levels Low to $High$ include all the states in $\mathcal{C}$. In other words, we set

$$Low = \min\{l \in \mathbb{Z}_+ \mid \mathcal{S}_l \cap \mathcal{C} \neq \emptyset\} \quad \text{and} \quad High = \max\{l \in \mathbb{Z}_+ \mid \mathcal{S}_l \cap \mathcal{C} \neq \emptyset\},$$

and the finite set $\bigcup_{Low \leq l \leq High} \mathcal{S}_l$ contains at least $(1 - \varepsilon)$ of the steady-state probability. We remark that $\mathcal{C} \subseteq \bigcup_{Low \leq l \leq High} \mathcal{S}_l$ due to the way in which Low and $High$ are defined, but this can only improve the quality of the bound.

Given the pair of level numbers, $(Low, High)$, the next step is to compute the matrices of conditional expected sojourn times for levels between Low and $High$ using (5.3). This requires us to determine a starting value for R_{High}. Since R_{High} can only be approximated when the state space is truncated, all computed matrices of conditional expected sojourn times between levels Low and $High$ will also be approximate. Clearly, the quality of the approximations increases as the steady-state probability mass concentrated on states between levels Low and $High$ approaches 1 (i.e., as ε approaches 0) [62]. Through a set of experiments, it is shown that setting R_{High} to 0 as suggested in [6] results in almost no loss of accuracy when ε is close to 0.

5.3 Case Study from Stochastic Chemical Kinetics

A *system of stochastic chemical kinetics* describes the dynamics of a number of molecules that interact through chemical reactions. The state space of the corresponding Markovian model considered here is discrete, countably infinite, and

H-dimensional [62]. Among the H state variables, H_I are countably infinite and represent molecule numbers, whereas H_F are finite and represent the finite control mechanism by which molecules interact. Since the finite control mechanism may or may not exist, it follows that $H_I \geq 2$, $H_F \geq 0$, and $H = H_I + H_F$. Without loss of generality, we assume that the first H_I state variable indices are assigned to the molecules. Hence, the state spaces of the variables satisfy $|\mathcal{S}^{(h)}| \to \infty$ for $h = 1, \ldots, H_I$ and $|\mathcal{S}^{(h)}| < \infty$ for $h = H_I + 1, \ldots, H$. Now, let $\bar{\mathcal{S}} \subseteq \times_{h=H_I+1}^{H} \mathcal{S}^{(h)}$ denote the set of states the finite variables can take.

Whereas it is the number of molecules that is relevant in chemical kinetics, in other domains it may be the number of species, genes, proteins, etc. Therefore, we will be using these terms interchangeably. The discrete Markovian model considered here is more realistic and shown to behave differently than the respective deterministic model, which is continuous and in the form of a system of ordinary differential equations [77, 97]. This is especially true when the numbers of different types of molecules are relatively small. In passing to the formal definition of transition classes we will be using, in many cases not all states in the product state space $\times_{h=1}^{H} \mathcal{S}^{(h)}$ are necessarily reachable. However, each state in the state space $\mathcal{S}$ is reachable from every other state since the associated CTMC is assumed to be irreducible.

A *transition class* k for $k = 1, \ldots, K$ is a pair

$$\left(\phi^{(k)} \prod_{h=1}^{H} f^{(k,h)}(i_h), \mathbf{v}^{(k)} \right),$$

where $\phi^{(k)} \in \mathbb{R}_{>0}$, $f^{(k,h)}(i_h) : \mathcal{S}^{(h)} \to \mathbb{R}_{\geq 0}$, and $\mathbf{v}^{(k)} \in \mathbb{Z}^{1 \times H}$ are respectively its *state-independent transition rate*, its *state-dependent transition rate* for variable i_h, $h = 1, \ldots, H$, and its *state-change vector* [115]. The first element of the pair,

$$\alpha_k(\mathbf{i}) := \phi^{(k)} \prod_{h=1}^{H} f^{(k,h)}(i_h),$$

specifies the *transition rate* from state $\mathbf{i} \in \mathcal{S}$ to state $(\mathbf{i} + \mathbf{v}^{(k)}) \in \mathcal{S}$. The second element of the pair, $\mathbf{v}^{(k)}$, specifies the successor state of the transition, where $v_h^{(k)}$ denotes the change in variable i_h due to a class k transition. Typically the number of reactions in a system of stochastic chemical kinetics is finite and each reaction corresponds to one transition class.

Example 4. Consider the model of a biological process of metabolite synthesis [104] with repressilator having the transition classes in Table 5.1. Here, $H = 6$, $H_I = 3$, $H_F = 3$, $\mathbf{i} = (i_1, i_2, i_3, i_4, i_5, i_6)$, $K = 12$, and $\lambda_1, \lambda_2, \lambda_3, \mu_1, \mu_2, \mu_3, \beta_0$, $\beta_1 \in \mathbb{R}_{>0}$. The system has three types of genes and three different control variables. The genes regulate each other's synthesis in a cyclic manner, type 1 genes regulating type 2 genes, type 2 genes regulating type 3 genes, and type 3 genes regulating type 1 genes. Each type of gene has a single binding site to which only one repressor at

Table 5.1 Transition classes of metabolite synthesis model with repressilator

k	$\phi^{(k)}$	$f^{(k,1)}(i_1)$	$f^{(k,2)}(i_2)$	$f^{(k,3)}(i_3)$	$f^{(k,4)}(i_4)$	$f^{(k,5)}(i_5)$	$f^{(k,6)}(i_6)$	$\mathbf{v}^{(k)}$
1	λ_1	1	1	1	1	1	$(1-i_6)$	$\mathbf{e}_1^T$
2	μ_1	i_1	1	1	1	1	1	$-\mathbf{e}_1^T$
3	β_0	i_1	1	1	$(1-i_4)$	1	1	$(-\mathbf{e}_1+\mathbf{e}_4)^T$
4	β_1	1	1	1	i_4	1	1	$(\mathbf{e}_1-\mathbf{e}_4)^T$
5	λ_2	1	1	1	$(1-i_4)$	1	1	$\mathbf{e}_2^T$
6	μ_2	1	i_2	1	1	1	1	$-\mathbf{e}_2^T$
7	β_0	1	i_2	1	1	$(1-i_5)$	1	$(-\mathbf{e}_2+\mathbf{e}_5)^T$
8	β_1	1	1	1	1	i_5	1	$(\mathbf{e}_2-\mathbf{e}_5)^T$
9	λ_3	1	1	1	1	$(1-i_5)$	1	$\mathbf{e}_3^T$
10	μ_3	1	1	i_3	1	1	1	$-\mathbf{e}_3^T$
11	β_0	1	1	i_3	1	1	$(1-i_6)$	$(-\mathbf{e}_3+\mathbf{e}_6)^T$
12	β_1	1	1	1	1	1	i_6	$(\mathbf{e}_3-\mathbf{e}_6)^T$

a time can bind. A gene regulates another type of gene by producing its own type of repressor, the repressor binding to the binding site of the gene to be regulated, and thereby repressing (or blocking) the other type of gene. This kind of binding, in which only one repressor can bind to a binding site and repress a type of gene, is said to be noncooperative. When the first control variable is set to 1 (i.e., $i_4 = 1$), a type 1 repressor is bound to the binding site of type 2 genes. Similarly, when $i_5 = 1$, a type 2 repressor is bound to the binding site of type 3 genes, and when $i_6 = 1$, a type 3 repressor is bound to the binding site of type 1 genes. When a control variable is set to 0 rather than 1, the corresponding binding site is unbound. Hence, we have $\mathcal{S}^{(1)} = \mathcal{S}^{(2)} = \mathcal{S}^{(3)} = \mathbb{Z}_+$, $\mathcal{S}^{(4)} = \mathcal{S}^{(5)} = \mathcal{S}^{(6)} = \{0,1\}$, $\bar{\mathcal{S}} = \mathcal{S}^{(4)} \times \mathcal{S}^{(5)} \times \mathcal{S}^{(6)}$, $|\bar{\mathcal{S}}| = 8$, and $\mathcal{S} = \mathcal{S}^{(1)} \times \mathcal{S}^{(2)} \times \mathcal{S}^{(3)} \times \bar{\mathcal{S}}$. Parameters λ_h and μ_h for $h = 1, 2, 3$ respectively denote state-independent production and degradation rates of type h genes, whereas parameters β_0 and β_1 respectively denote state-independent binding and unbinding rates. However, degradation and binding rates are in fact state dependent. In this example, $\mathcal{S}$ is equal to the product state space.

Having specified the parameters that form the blocks $Q_{l,l-1}$, $Q_{l,l}$, $Q_{l,l+1}$ at levels $l = Low, \ldots, High$, two more things need to be done to formulate their Kronecker representation. First, for each transition class the transition rate matrices associated with state variables must be specified, one matrix per state variable. Second, these transition rate matrices must be composed using the Kronecker product operator and the state-independent transition rates to form the blocks. It is these problems that we now consider.

We first associate transition rate matrices with countably infinite variables. Observe that there are H_l such transition rate matrices to be associated with each transition class. The transition rate matrix of countably infinite variable i_h for $h = 1, \ldots, H_l$ and transition class $k = 1, \ldots, K$ is denoted by $S_k^{(h)} \in \mathbb{R}_{\geq 0}^{|\mathcal{S}^{(h)}| \times |\mathcal{S}^{(h)}|}$ and given entrywise as

$$S_k^{(h)}(i_h, j_h) = \begin{cases} f^{(k,h)}(i_h) & \text{if } j_h = i_h + v_h^{(k)} \\ 0 & \text{otherwise} \end{cases} \quad \text{for } i_h, j_h \in \mathcal{S}^{(h)}.$$

Regarding finite variables when $H > H_I$, we prefer to define a combined transition rate matrix since we have observed in practice that $|\mathcal{S}^{(h)}|$ for $h = H_I + 1, \ldots, H$ is very small. Recalling that $\bar{\mathcal{S}}$ denotes the set of states finite variables can take and therefore $\bar{\mathcal{S}} \subseteq \times_{h=H_I+1}^{H} \mathcal{S}^{(h)}$, the combined transition rate matrix of finite variables for transition class $k = 1, \ldots, K$ is denoted by $\bar{S}_k \in \mathbb{R}_{\geq 0}^{|\bar{\mathcal{S}}| \times |\bar{\mathcal{S}}|}$ and is given entrywise as

$$\bar{S}_k((i_{H_I+1}, \ldots, i_H), (j_{H_I+1}, \ldots, j_H))$$
$$= \begin{cases} \prod_{h=H_I+1}^{H} f^{(k,h)}(i_h) & \text{if } (j_{H_I+1}, \ldots, j_H) = (i_{H_I+1}, \ldots, i_H) \\ & \qquad\qquad + (v_{H_I+1}^{(k)}, \ldots, v_H^{(k)}) \\ 0 & \text{otherwise} \end{cases}$$

$$\text{for } (i_{H_I+1}, \ldots, i_H), (j_{H_I+1}, \ldots, j_H) \in \bar{\mathcal{S}}.$$

When $H = H_I$, it is assumed that $|\bar{\mathcal{S}}| = 1$ and $\bar{S}_k = (1)$.

Example 4 (ctnd.). The transition rate matrices corresponding to the countably infinite variables of the model in Table 5.1 are obtained as

$$S_1^{(2)} = S_1^{(3)} = S_2^{(2)} = S_2^{(3)} = S_3^{(2)} = S_3^{(3)} = S_4^{(2)} = S_4^{(3)} = S_5^{(1)} = S_5^{(3)}$$
$$= S_6^{(1)} = S_6^{(3)} = S_7^{(1)} = S_7^{(3)} = S_8^{(1)} = S_8^{(3)} = S_9^{(1)} = S_9^{(2)} = S_{10}^{(1)} = S_{10}^{(2)}$$
$$= S_{11}^{(1)} = S_{11}^{(2)} = S_{12}^{(1)} = S_{12}^{(2)} = I_\infty,$$
$$S_1^{(1)} = S_4^{(1)} = S_5^{(2)} = S_8^{(2)} = S_9^{(3)} = S_{12}^{(3)} = \text{superdiag}((1, 1, \ldots)^T),$$
$$S_2^{(1)} = S_3^{(1)} = S_6^{(2)} = S_7^{(2)} = S_{10}^{(3)} = S_{11}^{(3)} = \text{subdiag}((1, 2, \ldots)^T).$$

The combined transition rate matrices corresponding to finite variables of the same model are obtained as

$$\bar{S}_1 = I_2 \otimes I_2 \otimes \text{diag}((1, 0)^T), \quad \bar{S}_2 = \bar{S}_6 = \bar{S}_{10} = I_2 \otimes I_2 \otimes I_2,$$
$$\bar{S}_3 = \text{superdiag}((1)^T) \otimes I_2 \otimes I_2, \quad \bar{S}_4 = \text{subdiag}((1)^T) \otimes I_2 \otimes I_2,$$
$$\bar{S}_5 = \text{diag}((1, 0)^T) \otimes I_2 \otimes I_2, \quad \bar{S}_7 = I_2 \otimes \text{superdiag}((1)^T) \otimes I_2,$$
$$\bar{S}_8 = I_2 \otimes \text{subdiag}((1)^T) \otimes I_2, \quad \bar{S}_9 = I_2 \otimes \text{diag}((1, 0)^T) \otimes I_2,$$
$$\bar{S}_{11} = I_2 \otimes I_2 \otimes \text{superdiag}((1)^T), \quad \bar{S}_{12} = I_2 \otimes I_2 \otimes \text{subdiag}((1)^T).$$

Our objective is to formulate a Kronecker representation for the nonzero blocks of Q from the transition rate matrices and the state-independent transition rates.

To this end, let us start by formally defining S_l as the subset of states corresponding to level $l \in \mathbb{Z}_+$ given by

$$S_l = \{\mathbf{i} \in S \mid \max_{h=1,\ldots,H_I} (i_h) = l\}, \quad S = \bigcup_{l=0}^{\infty} S_l, \quad S_l \cap S_u = \emptyset \text{ for } l \neq u.$$

The maximum function is justified by observing that the maximum valued variable among $i_1, \ldots, i_{H_I}$ in any state $\mathbf{i} \in S$ changes by at most one through any transition due to the particular form of the state-change vectors $\mathbf{v}^{(k)}$ in the transition classes for systems of stochastic chemical kinetics.

For each level l, the values a variable can take depend on the values of other variables. Therefore, first we define a partition of the values a countably infinite variable can take where there is no such dependency in a way similar to HMMs [19]. Then we introduce a partition of S_l based on the partitions of countably infinite variables defined previously. Let

$$S_{l,p}^{(h)} = \begin{cases} \{i_h \mid 0 \leq i_h \leq l-1\} & \text{if } h < p, \\ \{l\} & \text{if } h = p, \\ \{i_h \mid 0 \leq i_h \leq l\} & \text{if } h > p, \end{cases} \quad \text{for } h, p = 1, \ldots, H_I.$$

Then partition $p = 1, \ldots, H_I$ of S_l, denoted by $S_{l,p}$, is given by

$$S_{l,p} = \left\{\mathbf{i} \in S_l \mid (i_1, \ldots, i_{H_I}) \in \times_{h=1}^{H_I} S_{l,p}^{(h)} \text{ and } (i_{H_I+1}, \ldots, i_H) \in \bar{S}\right\}.$$

Finally,

$$S_l = \bigcup_{p=1}^{H_I} S_{l,p}, \quad S_{l,p} \cap S_{l,w} = \emptyset \text{ for } p \neq w.$$

Without loss of generality, the partitions $S_{l,p}$ are assumed to be ordered increasingly within S_l according to partition index p.

The number of states within levels Low and $High$ is given by

$$N(Low, High) = \sum_{l=Low}^{High} |\bar{S}| \left((l+1)^{H_I} - (l)^{H_I}\right) = |\bar{S}| \left((High+1)^{H_I} - Low^{H_I}\right).$$

Observe that the number of states at level $l \in \mathbb{Z}_+$ is $O(l^{H_I-1})$.

Example 4 (ctnd.). In our case, we have

$$S_{l,1}^{(1)} = \{l\}, \quad S_{l,1}^{(2)} = S_{l,1}^{(3)} = \{0, \ldots, l\},$$

$$S_{l,2}^{(1)} = \{0, \ldots, l-1\}, \quad S_{l,2}^{(2)} = \{l\}, \quad S_{l,2}^{(3)} = \{0, \ldots, l\},$$

$$S_{l,3}^{(1)} = S_{l,3}^{(2)} = \{0, \ldots, l-1\}, \quad S_{l,3}^{(3)} = \{l\},$$

implying

$$\times_{h=1}^3 S_{0,1}^{(h)} = \{(0,0,0)\}, \quad \times_{h=1}^3 S_{0,2}^{(h)} = \times_{h=1}^3 S_{0,3}^{(h)} = \emptyset,$$

$$\times_{h=1}^3 S_{1,1}^{(h)} = \{(1,0,0), (1,0,1), (1,1,0), (1,1,1)\},$$

$$\times_{h=1}^3 S_{1,2}^{(h)} = \{(0,1,0), (0,1,1)\}, \quad \times_{h=1}^3 S_{1,3}^{(h)} = \{(0,0,1)\},$$

$$\times_{h=1}^3 S_{2,1}^{(h)} = \{(2,0,0), (2,0,1), (2,0,2), (2,1,0), (2,1,1), (2,1,2),$$
$$(2,2,0), (2,2,1), (2,2,2)\},$$

$$\times_{h=1}^3 S_{2,2}^{(h)} = \{(0,2,0), (0,2,1), (0,2,2), (1,2,0), (1,2,1), (1,2,2)\},$$

$$\times_{h=1}^3 S_{2,3}^{(h)} = \{(0,0,2), (0,1,2), (1,0,2), (1,1,2)\},$$

and so on. Since

$$\bar{S} = \{(0,0,0), \ldots, (1,1,1)\},$$

we obtain

$$S_{l,1} = \left(\times_{h=1}^3 S_{l,1}^{(h)}\right) \times \bar{S} = \{(l,0,0,0,0,0), \ldots, (l,l,l,1,1,1)\},$$

$$S_{l,2} = \left(\times_{h=1}^3 S_{l,2}^{(h)}\right) \times \bar{S} = \{(0,l,0,0,0,0), \ldots, (l-1,l,l,1,1,1)\},$$

$$S_{l,3} = \left(\times_{h=1}^3 S_{l,3}^{(h)}\right) \times \bar{S} = \{(0,0,l,0,0,0), \ldots, (l-1,l-1,l,1,1,1)\}.$$

Now we are in a position to introduce the Kronecker representation of nonzero blocks in Q following the partitions of the subset of states at each level. Nonzero blocks $Q_{0,0}$, $Q_{0,1}$, $Q_{1,0}$, and $Q_{l,m}$ for $l \in \mathbb{Z}_+$, $m = l-1, l, l+1$ are respectively (1×1), $(1 \times H_I)$, $(H_I \times 1)$, and $(H_I \times H_I)$ block matrices as in

$$Q_{0,0} = \left(Q_{0,0}^{(1,1)} \right), \quad Q_{0,1} = \left(Q_{0,1}^{(1,1)} \cdots Q_{0,1}^{(1,H_I)} \right), \quad Q_{1,0} = \begin{pmatrix} Q_{1,0}^{(1,1)} \\ \vdots \\ Q_{1,0}^{(H_I,1)} \end{pmatrix},$$

$$Q_{l,m} = \begin{pmatrix} Q_{l,m}^{(1,1)} & \cdots & Q_{l,m}^{(1,H_I)} \\ \vdots & \ddots & \vdots \\ Q_{l,m}^{(H_I,1)} & \cdots & Q_{l,m}^{(H_I,H_I)} \end{pmatrix}.$$

Furthermore, blocks of $Q_{l,m}$ can be written in terms of transition rate matrices and state-independent transition rates as in

$$Q_{l,m}^{(p,w)} = \begin{cases} \tilde{Q}_{l,m}^{(p,w)} - \mathrm{diag}\left(\sum_{m'=l-1}^{l+1} \sum_{w'=1}^{H_I} \tilde{Q}_{l,m'}^{(p,w')} \mathbf{e}\right) & \text{if } p = w \text{ and } l = m \\ \tilde{Q}_{l,m}^{(p,w)} & \text{otherwise} \end{cases}$$

for $l \in \mathbb{Z}_+$, $m = l - 1, l, l + 1$ and $p, w = 1, \ldots, H_I$, where

$$\tilde{Q}_{l,m}^{(p,w)} = \sum_{k=1}^{K} \phi^{(k)} \left(\bigotimes_{h=1}^{H_I} S_k^{(h)}(\mathcal{S}_{l,p}^{(h)}, \mathcal{S}_{m,w}^{(h)})\right) \otimes \bar{S}_k$$

and $S_k^{(h)}(\mathcal{S}_{l,p}^{(h)}, \mathcal{S}_{m,w}^{(h)})$ denotes the submatrix of $S_k^{(h)}$ incident on row indices in $\mathcal{S}_{l,p}^{(h)}$ and column indices in $\mathcal{S}_{m,w}^{(h)}$. The first summation in diag for $Q_{l,m}^{(p,w)}$ should have a starting index of 0 rather than -1 for the equation of block $Q_{0,0}^{(1,1)}$. Also, the ending index of the second summation in diag should be 1 rather than H_I for $Q_{1,1}^{(1,1)}, \ldots, Q_{1,1}^{(H_I,H_I)}$ when $m' = l - 1$. Observe that beyond the boundary level, we have nonzero $(N \times N)$ block (sub)matrices along the block tridiagonal structure, where $N = H_I$, as in Sect. 4.2.

Example 4 (cntd.). Since $H_I = 3$, the nonzero blocks $Q_{0,0}$, $Q_{0,1}$, $Q_{1,0}$, and $Q_{l,m}$ for $l \in \mathbb{Z}_+$, $m = l - 1, l, l + 1$ (except $Q_{1,0}$) are respectively (1×1), (1×3), (3×1), and (3×3) block matrices. In particular, the seven blocks associated with $Q_{0,0}$, $Q_{0,1}$, $Q_{1,0}$ are given by

$$\tilde{Q}_{0,0}^{(1,1)} = \sum_{k=1}^{12} \phi^{(k)} \left(\bigotimes_{h=1}^{3} S_k^{(h)}(\mathcal{S}_{0,1}^{(h)}, \mathcal{S}_{0,1}^{(h)})\right) \otimes \bar{S}_k = (0)_{8\times 8},$$

$$\tilde{Q}_{0,1}^{(1,1)} = \sum_{k=1}^{12} \phi^{(k)} \left(\bigotimes_{h=1}^{3} S_k^{(h)}(\mathcal{S}_{0,1}^{(h)}, \mathcal{S}_{1,1}^{(h)})\right) \otimes \bar{S}_k$$
$$= \lambda_1 (1) \otimes (1,0) \otimes (1,0) \otimes \bar{S}_1 + \beta_1 (1) \otimes (1,0) \otimes (1,0) \otimes \bar{S}_4,$$

$$\tilde{Q}_{0,1}^{(1,2)} = \sum_{k=1}^{12} \phi^{(k)} \left(\bigotimes_{h=1}^{3} S_k^{(h)}(\mathcal{S}_{0,1}^{(h)}, \mathcal{S}_{1,2}^{(h)})\right) \otimes \bar{S}_k$$
$$= \lambda_2 (1) \otimes (1) \otimes (1,0) \otimes \bar{S}_5 + \beta_1 (1) \otimes (1) \otimes (1,0) \otimes \bar{S}_8,$$

$$\tilde{Q}_{0,1}^{(1,3)} = \sum_{k=1}^{12} \phi^{(k)} \left(\bigotimes_{h=1}^{3} S_k^{(h)}(\mathcal{S}_{0,1}^{(h)}, \mathcal{S}_{1,3}^{(h)})\right) \otimes \bar{S}_k$$
$$= \lambda_3 (1) \otimes (1) \otimes (1) \otimes \bar{S}_9 + \beta_1 (1) \otimes (1) \otimes (1) \otimes \bar{S}_{12},$$

$$\tilde{Q}_{1,0}^{(1,1)} = \sum_{k=1}^{12} \phi^{(k)} \left(\bigotimes_{h=1}^{3} S_k^{(h)}(\mathcal{S}_{1,1}^{(h)}, \mathcal{S}_{0,1}^{(h)})\right) \otimes \bar{S}_k$$
$$= \mu_1 (1) \otimes (1,0)^T \otimes (1,0)^T \otimes \bar{S}_2 + \beta_0 (1) \otimes (1,0)^T \otimes (1,0)^T \otimes \bar{S}_3,$$

$$\tilde{Q}_{1,0}^{(2,1)} = \sum_{k=1}^{12} \phi^{(k)} \left(\bigotimes_{h=1}^{3} S_k^{(h)}(S_{1,2}^{(h)}, S_{0,1}^{(h)}) \right) \otimes \bar{S}_k$$

$$= \mu_2 \, (1) \otimes (1) \otimes (1,0)^T \otimes \bar{S}_6 + \beta_0 \, (1) \otimes (1) \otimes (1,0)^T \otimes \bar{S}_7,$$

$$\tilde{Q}_{1,0}^{(3,1)} = \sum_{k=1}^{12} \phi^{(k)} \left(\bigotimes_{h=1}^{3} S_k^{(h)}(S_{1,3}^{(h)}, S_{0,1}^{(h)}) \right) \otimes \bar{S}_k$$

$$= \mu_3 \, (1) \otimes (1) \otimes (1) \otimes \bar{S}_{10} + \beta_0 \, (1) \otimes (1) \otimes (1) \otimes \bar{S}_{11},$$

the nine blocks associated with $Q_{l,l-1}$ are given by

$$\tilde{Q}_{l,l-1}^{(1,1)} = \sum_{k=1}^{12} \phi^{(k)} \left(\bigotimes_{h=1}^{3} S_k^{(h)}(S_{l,1}^{(h)}, S_{l-1,1}^{(h)}) \right) \otimes \bar{S}_k$$

$$= \mu_1 \, (l) \otimes \mathrm{diag}(\mathbf{e})_{(l+1)\times l} \otimes \mathrm{diag}(\mathbf{e})_{(l+1)\times l} \otimes \bar{S}_2$$

$$+ \beta_0 \, (l) \otimes \mathrm{diag}(\mathbf{e})_{(l+1)\times l} \otimes \mathrm{diag}(\mathbf{e})_{(l+1)\times l} \otimes \bar{S}_3,$$

$$\tilde{Q}_{l,l-1}^{(1,2)} = \sum_{k=1}^{12} \phi^{(k)} \left(\bigotimes_{h=1}^{3} S_k^{(h)}(S_{l,1}^{(h)}, S_{l-1,2}^{(h)}) \right) \otimes \bar{S}_k = 0_{8(l+1)^2 \times 8(l-1)l},$$

$$\tilde{Q}_{l,l-1}^{(1,3)} = \sum_{k=1}^{12} \phi^{(k)} \left(\bigotimes_{h=1}^{3} S_k^{(h)}(S_{l,1}^{(h)}, S_{l-1,3}^{(h)}) \right) \otimes \bar{S}_k = 0_{8(l+1)^2 \times 8(l-1)^2},$$

$$\tilde{Q}_{l,l-1}^{(2,1)} = \sum_{k=1}^{12} \phi^{(k)} \left(\bigotimes_{h=1}^{3} S_k^{(h)}(S_{l,2}^{(h)}, S_{l-1,1}^{(h)}) \right) \otimes \bar{S}_k$$

$$= \mu_2 \, (\mathbf{e}_l)_{l\times 1} \otimes (l \, \mathbf{e}_l^T)_{1\times l} \otimes \mathrm{diag}(\mathbf{e})_{(l+1)\times l} \otimes \bar{S}_6$$

$$+ \beta_0 \, (\mathbf{e}_l)_{l\times 1} \otimes (l \, \mathbf{e}_l^T)_{1\times l} \otimes \mathrm{diag}(\mathbf{e})_{(l+1)\times l} \otimes \bar{S}_7,$$

$$\tilde{Q}_{l,l-1}^{(2,2)} = \sum_{k=1}^{12} \phi^{(k)} \left(\bigotimes_{h=1}^{3} S_k^{(h)}(S_{l,2}^{(h)}, S_{l-1,2}^{(h)}) \right) \otimes \bar{S}_k$$

$$= \mu_2 \, \mathrm{diag}(\mathbf{e})_{l\times(l-1)} \otimes (l) \otimes \mathrm{diag}(\mathbf{e})_{(l+1)\times l} \otimes \bar{S}_6$$

$$+ \beta_0 \, \mathrm{diag}(\mathbf{e})_{l\times(l-1)} \otimes (l) \otimes \mathrm{diag}(\mathbf{e})_{(l+1)\times l} \otimes \bar{S}_7,$$

$$\tilde{Q}_{l,l-1}^{(2,3)} = \sum_{k=1}^{12} \phi^{(k)} \left(\bigotimes_{h=1}^{3} S_k^{(h)}(S_{l,2}^{(h)}, S_{l-1,3}^{(h)}) \right) \otimes \bar{S}_k = 0_{8l(l+1)\times 8(l-1)^2},$$

$$\tilde{Q}_{l,l-1}^{(3,1)} = \sum_{k=1}^{12} \phi^{(k)} \left(\bigotimes_{h=1}^{3} S_k^{(h)}(S_{l,3}^{(h)}, S_{l-1,1}^{(h)}) \right) \otimes \bar{S}_k$$

$$= \mu_3 \, (\mathbf{e}_l)_{l\times 1} \otimes \mathrm{diag}(\mathbf{e})_{l\times l} \otimes (l \, \mathbf{e}_l^T)_{1\times l} \otimes \bar{S}_{10}$$

$$+ \beta_0 \, (\mathbf{e}_l)_{l\times 1} \otimes \mathrm{diag}(\mathbf{e})_{l\times l} \otimes (l \, \mathbf{e}_l^T)_{1\times l} \otimes \bar{S}_{11},$$

$$\tilde{Q}_{l,l-1}^{(3,2)} = \sum_{k=1}^{12} \phi^{(k)} \left(\bigotimes_{h=1}^{3} S_k^{(h)}(\mathcal{S}_{l,3}^{(h)}, \mathcal{S}_{l-1,2}^{(h)}) \right) \otimes \bar{S}_k$$

$$= \mu_3 \, \text{diag}(\mathbf{e})_{l \times (l-1)} \otimes (\mathbf{e}_l)_{l \times 1} \otimes (l\mathbf{e}_l^T)_{1 \times l} \otimes \bar{S}_{10}$$

$$+ \beta_0 \, \text{diag}(\mathbf{e})_{l \times (l-1)} \otimes (\mathbf{e}_l)_{l \times 1} \otimes (l\mathbf{e}_l^T)_{1 \times l} \otimes \bar{S}_{11},$$

$$\tilde{Q}_{l,l-1}^{(3,3)} = \sum_{k=1}^{12} \phi^{(k)} \left(\bigotimes_{h=1}^{3} S_k^{(h)}(\mathcal{S}_{l,3}^{(h)}, \mathcal{S}_{l-1,3}^{(h)}) \right) \otimes \bar{S}_k$$

$$= \mu_3 \, \text{diag}(\mathbf{e})_{l \times (l-1)} \otimes \text{diag}(\mathbf{e})_{l \times (l-1)} \otimes (l) \otimes \bar{S}_{10}$$

$$+ \beta_0 \, \text{diag}(\mathbf{e})_{l \times (l-1)} \otimes \text{diag}(\mathbf{e})_{l \times (l-1)} \otimes (l) \otimes \bar{S}_{11},$$

the nine blocks associated with $Q_{l,l}$ are given by

$$\tilde{Q}_{l,l}^{(1,1)} = \sum_{k=1}^{12} \phi^{(k)} \left(\bigotimes_{h=1}^{3} S_k^{(h)}(\mathcal{S}_{l,1}^{(h)}, \mathcal{S}_{l,1}^{(h)}) \right) \otimes \bar{S}_k$$

$$= \lambda_2 \, (1) \otimes \text{superdiag}(\mathbf{e})_{(l+1) \times (l+1)} \otimes \text{diag}(\mathbf{e})_{(l+1) \times (l+1)} \otimes \bar{S}_5$$

$$+ \mu_2 \, (1) \otimes \text{subdiag}((1, \ldots, l)^T)_{(l+1) \times (l+1)} \otimes \text{diag}(\mathbf{e})_{(l+1) \times (l+1)} \otimes \bar{S}_6$$

$$+ \beta_0 \, (1) \otimes \text{subdiag}((1, \ldots, l)^T)_{(l+1) \times (l+1)} \otimes \text{diag}(\mathbf{e})_{(l+1) \times (l+1)} \otimes \bar{S}_7$$

$$+ \beta_1 \, (1) \otimes \text{superdiag}(\mathbf{e})_{(l+1) \times (l+1)} \otimes \text{diag}(\mathbf{e})_{(l+1) \times (l+1)} \otimes \bar{S}_8$$

$$+ \lambda_3 \, (1) \otimes \text{diag}(\mathbf{e})_{(l+1) \times (l+1)} \otimes \text{superdiag}(\mathbf{e})_{(l+1) \times (l+1)} \otimes \bar{S}_9$$

$$+ \mu_3 \, (1) \otimes \text{diag}(\mathbf{e})_{(l+1) \times (l+1)} \otimes \text{subdiag}((1, \ldots, l)^T)_{(l+1) \times (l+1)} \otimes \bar{S}_{10}$$

$$+ \beta_0 \, (1) \otimes \text{diag}(\mathbf{e})_{(l+1) \times (l+1)} \otimes \text{subdiag}((1, \ldots, l)^T)_{(l+1) \times (l+1)} \otimes \bar{S}_{11}$$

$$+ \beta_1 \, (1) \otimes \text{diag}(\mathbf{e})_{(l+1) \times (l+1)} \otimes \text{superdiag}(\mathbf{e})_{(l+1) \times (l+1)} \otimes \bar{S}_{12},$$

$$\tilde{Q}_{l,l}^{(1,2)} = \sum_{k=1}^{12} \phi^{(k)} \left(\bigotimes_{h=1}^{3} S_k^{(h)}(\mathcal{S}_{l,1}^{(h)}, \mathcal{S}_{l,2}^{(h)}) \right) \otimes \bar{S}_k$$

$$= \mu_1 \, (l\mathbf{e}_l^T)_{1 \times l} \otimes (\mathbf{e}_{l+1})_{(l+1) \times 1} \otimes \text{diag}(\mathbf{e})_{(l+1) \times (l+1)} \otimes \bar{S}_2$$

$$+ \beta_0 \, (l\mathbf{e}_l^T)_{1 \times l} \otimes (\mathbf{e}_{l+1})_{(l+1) \times 1} \otimes \text{diag}(\mathbf{e})_{(l+1) \times (l+1)} \otimes \bar{S}_3,$$

$$\tilde{Q}_{l,l}^{(1,3)} = \sum_{k=1}^{12} \phi^{(k)} \left(\bigotimes_{h=1}^{3} S_k^{(h)}(\mathcal{S}_{l,1}^{(h)}, \mathcal{S}_{l,3}^{(h)}) \right) \otimes \bar{S}_k$$

$$= \mu_1 \, (l\mathbf{e}_l^T)_{1 \times l} \otimes \text{diag}(\mathbf{e})_{(l+1) \times l} \otimes (\mathbf{e}_{l+1})_{(l+1) \times 1} \otimes \bar{S}_2$$

$$+ \beta_0 \, (l\mathbf{e}_l^T)_{1 \times l} \otimes \text{diag}(\mathbf{e})_{(l+1) \times l} \otimes (\mathbf{e}_{l+1})_{(l+1) \times 1} \otimes \bar{S}_3,$$

$$\tilde{Q}_{l,l}^{(2,1)} = \sum_{k=1}^{12} \phi^{(k)} \left(\bigotimes_{h=1}^{3} S_k^{(h)}(\mathcal{S}_{l,2}^{(h)}, \mathcal{S}_{l,1}^{(h)}) \right) \otimes \bar{S}_k$$

$$= \lambda_1 \, (\mathbf{e}_l)_{l\times 1} \otimes (\mathbf{e}_{l+1}^T)_{1\times(l+1)} \otimes \mathrm{diag}(\mathbf{e})_{(l+1)\times(l+1)} \otimes \bar{S}_1$$

$$+ \beta_1 \, (\mathbf{e}_l)_{l\times 1} \otimes (\mathbf{e}_{l+1}^T)_{1\times(l+1)} \otimes \mathrm{diag}(\mathbf{e})_{(l+1)\times(l+1)} \otimes \bar{S}_4,$$

$$\tilde{Q}_{l,l}^{(2,2)} = \sum_{k=1}^{12} \phi^{(k)} \left(\bigotimes_{h=1}^{3} S_k^{(h)}(\mathcal{S}_{l,2}^{(h)}, \mathcal{S}_{l,2}^{(h)}) \right) \otimes \bar{S}_k$$

$$= \lambda_1 \, \mathrm{superdiag}(\mathbf{e})_{l\times l} \otimes (1) \otimes \mathrm{diag}(\mathbf{e})_{(l+1)\times(l+1)} \otimes \bar{S}_1$$

$$+ \mu_1 \, \mathrm{subdiag}((1,\ldots,l-1)^T)_{l\times l} \otimes (1) \otimes \mathrm{diag}(\mathbf{e})_{(l+1)\times(l+1)} \otimes \bar{S}_2$$

$$+ \beta_0 \, \mathrm{subdiag}((1,\ldots,l-1)^T)_{l\times l} \otimes (1) \otimes \mathrm{diag}(\mathbf{e})_{(l+1)\times(l+1)} \otimes \bar{S}_3$$

$$+ \beta_1 \, \mathrm{superdiag}(\mathbf{e})_{l\times l} \otimes (1) \otimes \mathrm{diag}(\mathbf{e})_{(l+1)\times(l+1)} \otimes \bar{S}_4$$

$$+ \lambda_3 \, \mathrm{diag}(\mathbf{e})_{l\times l} \otimes (1) \otimes \mathrm{superdiag}(\mathbf{e})_{(l+1)\times(l+1)} \otimes \bar{S}_9$$

$$+ \mu_3 \, \mathrm{diag}(\mathbf{e})_{l\times l} \otimes (1) \otimes \mathrm{subdiag}((1,\ldots,l)^T)_{(l+1)\times(l+1)} \otimes \bar{S}_{10}$$

$$+ \beta_0 \, \mathrm{diag}(\mathbf{e})_{l\times l} \otimes (1) \otimes \mathrm{subdiag}((1,\ldots,l)^T)_{(l+1)\times(l+1)} \otimes \bar{S}_{11}$$

$$+ \beta_1 \, \mathrm{diag}(\mathbf{e})_{l\times l} \otimes (1) \otimes \mathrm{superdiag}(\mathbf{e})_{(l+1)\times(l+1)} \otimes \bar{S}_{12},$$

$$\tilde{Q}_{l,l}^{(2,3)} = \sum_{k=1}^{12} \phi^{(k)} \left(\bigotimes_{h=1}^{3} S_k^{(h)}(\mathcal{S}_{l,2}^{(h)}, \mathcal{S}_{l,3}^{(h)}) \right) \otimes \bar{S}_k$$

$$= \mu_2 \, \mathrm{diag}(\mathbf{e})_{l\times l} \otimes (l\mathbf{e}_l^T)_{1\times l} \otimes (\mathbf{e}_{l+1})_{(l+1)\times 1} \otimes \bar{S}_6$$

$$+ \beta_0 \, \mathrm{diag}(\mathbf{e})_{l\times l} \otimes (l\mathbf{e}_l^T)_{1\times l} \otimes (\mathbf{e}_{l+1})_{(l+1)\times 1} \otimes \bar{S}_7,$$

$$\tilde{Q}_{l,l}^{(3,1)} = \sum_{k=1}^{12} \phi^{(k)} \left(\bigotimes_{h=1}^{3} S_k^{(h)}(\mathcal{S}_{l,3}^{(h)}, \mathcal{S}_{l,1}^{(h)}) \right) \otimes \bar{S}_k$$

$$= \lambda_1 \, (\mathbf{e}_l)_{l\times 1} \otimes \mathrm{diag}(\mathbf{e})_{l\times(l+1)} \otimes (\mathbf{e}_{l+1}^T)_{1\times(l+1)} \otimes \bar{S}_1$$

$$+ \beta_1 \, (\mathbf{e}_l)_{l\times 1} \otimes \mathrm{diag}(\mathbf{e})_{l\times(l+1)} \otimes (\mathbf{e}_{l+1}^T)_{1\times(l+1)} \otimes \bar{S}_4,$$

$$\tilde{Q}_{l,l}^{(3,2)} = \sum_{k=1}^{12} \phi^{(k)} \left(\bigotimes_{h=1}^{3} S_k^{(h)}(\mathcal{S}_{l,3}^{(h)}, \mathcal{S}_{l,2}^{(h)}) \right) \otimes \bar{S}_k$$

$$= \lambda_2 \, \mathrm{diag}(\mathbf{e})_{l\times l} \otimes (\mathbf{e}_l)_{l\times 1} \otimes (\mathbf{e}_{l+1}^T)_{1\times(l+1)} \otimes \bar{S}_5$$

$$+ \beta_1 \, \mathrm{diag}(\mathbf{e})_{l\times l} \otimes (\mathbf{e}_l)_{l\times 1} \otimes (\mathbf{e}_{l+1}^T)_{1\times(l+1)} \otimes \bar{S}_8,$$

$$\tilde{Q}_{l,l}^{(3,3)} = \sum_{k=1}^{12} \phi^{(k)} \left(\bigotimes_{h=1}^{3} S_k^{(h)}(\mathcal{S}_{l,3}^{(h)}, \mathcal{S}_{l,3}^{(h)}) \right) \otimes \bar{S}_k$$

$$= \lambda_1 \; \text{superdiag}(\mathbf{e})_{l \times l} \otimes \text{diag}(\mathbf{e})_{l \times l} \otimes (1) \otimes \bar{S}_1$$

$$+ \mu_1 \; \text{subdiag}((1, \ldots, l-1)^T)_{l \times l} \otimes \text{diag}(\mathbf{e})_{l \times l} \otimes (1) \otimes \bar{S}_2$$

$$+ \beta_0 \; \text{subdiag}((1, \ldots, l-1)^T)_{l \times l} \otimes \text{diag}(\mathbf{e})_{l \times l} \otimes (1) \otimes \bar{S}_3$$

$$+ \beta_1 \; \text{superdiag}(\mathbf{e})_{l \times l} \otimes \text{diag}(\mathbf{e})_{l \times l} \otimes (1) \otimes \bar{S}_4$$

$$+ \lambda_2 \; \text{diag}(\mathbf{e})_{l \times l} \otimes \text{superdiag}(\mathbf{e})_{l \times l} \otimes (1) \otimes \bar{S}_5$$

$$+ \mu_2 \; \text{diag}(\mathbf{e})_{l \times l} \otimes \text{subdiag}((1, \ldots, l-1)^T)_{l \times l} \otimes (1) \otimes \bar{S}_6$$

$$+ \beta_0 \; \text{diag}(\mathbf{e})_{l \times l} \otimes \text{subdiag}((1, \ldots, l-1)^T)_{l \times l} \otimes (1) \otimes \bar{S}_7$$

$$+ \beta_1 \; \text{diag}(\mathbf{e})_{l \times l} \otimes \text{superdiag}(\mathbf{e})_{l \times l} \otimes (1) \otimes \bar{S}_8,$$

and the nine blocks associated with $Q_{l,l+1}$ are given by

$$\tilde{Q}_{l,l+1}^{(1,1)} = \sum_{k=1}^{12} \phi^{(k)} \left(\bigotimes_{h=1}^{3} S_k^{(h)}(\mathcal{S}_{l,1}^{(h)}, \mathcal{S}_{l+1,1}^{(h)}) \right) \otimes \bar{S}_k$$

$$= \lambda_1 \; (1) \otimes \text{diag}(\mathbf{e})_{(l+1) \times (l+2)} \otimes \text{diag}(\mathbf{e})_{(l+1) \times (l+2)} \otimes \bar{S}_1$$

$$+ \beta_1 \; (1) \otimes \text{diag}(\mathbf{e})_{(l+1) \times (l+2)} \otimes \text{diag}(\mathbf{e})_{(l+1) \times (l+2)} \otimes \bar{S}_4,$$

$$\tilde{Q}_{l,l+1}^{(1,2)} = \sum_{k=1}^{12} \phi^{(k)} \left(\bigotimes_{h=1}^{3} S_k^{(h)}(\mathcal{S}_{l,1}^{(h)}, \mathcal{S}_{l+1,2}^{(h)}) \right) \otimes \bar{S}_k$$

$$= \lambda_2 \; (\mathbf{e}_{l+1}^T)_{1 \times (l+1)} \otimes (\mathbf{e}_{l+1})_{(l+1) \times 1} \otimes \text{diag}(\mathbf{e})_{(l+1) \times (l+2)} \otimes \bar{S}_5$$

$$+ \beta_1 \; (\mathbf{e}_{l+1}^T)_{1 \times (l+1)} \otimes (\mathbf{e}_{l+1})_{(l+1) \times 1} \otimes \text{diag}(\mathbf{e})_{(l+1) \times (l+2)} \otimes \bar{S}_8,$$

$$\tilde{Q}_{l,l+1}^{(1,3)} = \sum_{k=1}^{12} \phi^{(k)} \left(\bigotimes_{h=1}^{3} S_k^{(h)}(\mathcal{S}_{l,1}^{(h)}, \mathcal{S}_{l+1,3}^{(h)}) \right) \otimes \bar{S}_k$$

$$= \lambda_3 \; (\mathbf{e}_{l+1}^T)_{1 \times (l+1)} \otimes \text{diag}(\mathbf{e})_{(l+1) \times (l+1)} \otimes (\mathbf{e}_{l+1})_{(l+1) \times 1} \otimes \bar{S}_9$$

$$+ \beta_1 \; (\mathbf{e}_{l+1}^T)_{1 \times (l+1)} \otimes \text{diag}(\mathbf{e})_{(l+1) \times (l+1)} \otimes (\mathbf{e}_{l+1})_{(l+1) \times 1}^T \otimes \bar{S}_{12},$$

$$\tilde{Q}_{l,l+1}^{(2,1)} = \sum_{k=1}^{12} \phi^{(k)} \left(\bigotimes_{h=1}^{3} S_k^{(h)}(\mathcal{S}_{l,2}^{(h)}, \mathcal{S}_{l+1,1}^{(h)}) \right) \otimes \bar{S}_k = 0_{8l(l+1) \times 8(l+2)^2},$$

$$\tilde{Q}_{l,l+1}^{(2,2)} = \sum_{k=1}^{12} \phi^{(k)} \left(\bigotimes_{h=1}^{3} S_k^{(h)}(\mathcal{S}_{l,2}^{(h)}, \mathcal{S}_{l+1,2}^{(h)}) \right) \otimes \bar{S}_k$$

$$= \lambda_2 \; \text{diag}(\mathbf{e})_{l \times (l+1)} \otimes (1) \otimes \text{diag}(\mathbf{e})_{(l+1) \times (l+2)} \otimes \bar{S}_5$$

$$+ \beta_1 \; \text{diag}(\mathbf{e})_{l \times (l+1)} \otimes (1) \otimes \text{diag}(\mathbf{e})_{(l+1) \times (l+2)} \otimes \bar{S}_8,$$

$$\tilde{Q}_{l,l+1}^{(2,3)} = \sum_{k=1}^{12} \phi^{(k)} \left(\bigotimes_{h=1}^{3} S_k^{(h)}(\mathcal{S}_{l,2}^{(h)}, \mathcal{S}_{l+1,3}^{(h)}) \right) \otimes \bar{S}_k$$

$$= \lambda_3 \, \mathrm{diag}(\mathbf{e})_{l \times (l+1)} \otimes (\mathbf{e}_{l+1}^T)_{1 \times (l+1)} \otimes (\mathbf{e}_{l+1})_{(l+1) \times 1} \otimes \bar{S}_9$$

$$+ \beta_1 \, \mathrm{diag}(\mathbf{e})_{l \times (l+1)} \otimes (\mathbf{e}_{l+1}^T)_{1 \times (l+1)} \otimes (\mathbf{e}_{l+1})_{(l+1) \times 1} \otimes \bar{S}_{12},$$

$$\tilde{Q}_{l,l+1}^{(3,1)} = \sum_{k=1}^{12} \phi^{(k)} \left(\bigotimes_{h=1}^{3} S_k^{(h)}(\mathcal{S}_{l,3}^{(h)}, \mathcal{S}_{l+1,1}^{(h)}) \right) \otimes \bar{S}_k = 0_{8l^2 \times 8(l+2)^2},$$

$$\tilde{Q}_{l,l+1}^{(3,2)} = \sum_{k=1}^{12} \phi^{(k)} \left(\bigotimes_{h=1}^{3} S_k^{(h)}(\mathcal{S}_{l,3}^{(h)}, \mathcal{S}_{l+1,2}^{(h)}) \right) \otimes \bar{S}_k = 0_{8l^2 \times 8(l+1)(l+2)},$$

$$\tilde{Q}_{l,l+1}^{(3,3)} = \sum_{k=1}^{12} \phi^{(k)} \left(\bigotimes_{h=1}^{3} S_k^{(h)}(\mathcal{S}_{l,3}^{(h)}, \mathcal{S}_{l+1,3}^{(h)}) \right) \otimes \bar{S}_k$$

$$= \lambda_3 \, \mathrm{diag}(\mathbf{e})_{l \times (l+1)} \otimes \mathrm{diag}(\mathbf{e})_{l \times (l+1)} \otimes (1) \otimes \bar{S}_9$$

$$+ \beta_1 \, \mathrm{diag}(\mathbf{e})_{l \times (l+1)} \otimes \mathrm{diag}(\mathbf{e})_{l \times (l+1)} \otimes (1) \otimes \bar{S}_{12}.$$

Example 4 (ctnd.). Now let us investigate the suitability of the squared Euclidean norm, that is, $g(i_1, i_2, i_3, i_4, i_5, i_6) = \sum_{h=1}^{6} i_h^2$, as the Lyapunov function for the set of parameters $\lambda_1 = \lambda_2 = \lambda_3 = 1.3$, $\mu_1 = \mu_2 = \mu_3 = 0.8$, $\beta_0 = 1$, and $\beta_1 = 0.5$. The corresponding drift from (5.4) is given by

$$d(i_1, i_2, i_3, i_4, i_5, i_6) = -3.6i_1^2 + 2i_1^2 i_4 - i_1 i_4 - 2.6i_1 i_6 + 5.4i_1 - 1.3i_6 + 1.3$$

$$-3.6i_2^2 + 2i_2^2 i_5 - i_2 i_5 - 2.6i_2 i_4 + 5.4i_2 - 1.3i_4 + 1.3$$

$$-3.6i_3^2 + 2i_3^2 i_6 - i_3 i_6 - 2.6i_3 i_5 + 5.4i_3 - 1.3i_5 + 1.3.$$

Observe that $d(i_1, i_2, i_3, i_4, i_5, i_6)$ is a nonlinear function of six variables. The proof that $\mathcal{C}$ is finite follows from showing that there exists a finite superset of $\mathcal{C}$ [115]. The global maximum drift is computed as $c = 9.3$ using the HOM4PS2-2.0 package [102]. It is attained at state $(1, 1, 1, 0, 0, 0)$.

With a lower bound of 95% on the steady-state probability mass (i.e., $\varepsilon = 0.05$), we obtain $\gamma = 176.7$. This yields $(Low, High) = (0, 12)$ and an infinity residual norm, $\|\tilde{\pi} Q\|_\infty$, on the order of 10^{-8} in about 133 s using 335 MB on a PC with an Intel Core2 Duo 1.83-GHz processor and 4 GB of main memory. The code is available in Matlab [57]. The LDQBD model restricted to levels $(0, 12)$ has 17,576 states. Note that the computation of the residual norm is restricted to these states since all other entries in π are taken to be zero, hence the notation $\tilde{\pi}$ for an approximation of π. Given more memory and time, it is always possible to

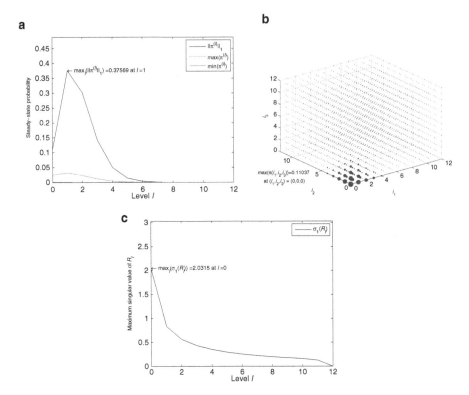

Fig. 5.1 Plots for repressilator model with $(\lambda_1, \lambda_2, \lambda_3, \mu_1, \mu_2, \mu_3, \beta_0, \beta_1) = (1.3, 1.3, 1.3, 0.8,$ $0.8, 0.8, 1, 0.5)$ and $(Low, High) = (0, 12)$. **a** $||\tilde{\pi}^{(l)}||_1$, $\max(\tilde{\pi}^{(l)})$, and $\min(\tilde{\pi}^{(l)})$ across levels. **b** $\tilde{\pi}(i_1, i_2, i_3)$ across countably infinite state variables. **c** $\sigma_1(\tilde{R}_l)$ across levels

obtain more accurate results with the matrix-analytic approach [62]. For instance, $High = 13$ yields a residual norm of 10^{-9} and $High = 14$ yields a residual norm of 10^{-10}, though both were obtained in an increasing amount of time and with larger memory consumption.

The marginal steady-state probability across levels is maximized at level 1 (Fig. 5.1a) and the marginal steady-state probability across the countably infinite variables is maximized at level 0 (Fig. 5.1b), where the size of a sphere indicates the magnitude of the corresponding marginal steady-state probability. We report results to five decimal digits of precision. Recall that the matrices of conditional expected sojourn times, R_l, are nonnegative and rectangular. For the problem at hand, these matrices have percentages of nonzeros roughly above 80%, implying they may very well be stored as full matrices. Such high percentages have also been observed in other problems [62]. The behavior of their largest singular values [78, 111], denoted by $\sigma_1(R_l)$, for $l = Low, \ldots, High$ provides insight regarding the behavior of π for the system under study. Recall that the singular values of a real rectangular matrix X are the square roots of the eigenvalues of $X^T X$, which by construction is symmetric positive definite and therefore has positive real eigenvalues. In our problem, the maximum singular value of R_l appears at level 0 in Fig. 5.1c. The expected values of the three countably infinite variables are obtained as 0.75701.

For comparison purposes, the same model is simulated with the benchmark implementation StochKit [103, 142]. For 31 sample paths amounting to a total of about 4×10^9 transitions in 761 s, the simulation yields the expected values of the three countably infinite variables with confidence intervals corresponding to a probability of 0.95 as 0.75708 ± 0.00023, 0.75708 ± 0.00019, and 0.75699 ± 0.00024, respectively. Observe that there are about four to five correct digits in these results.

In [61], an alternative technique for systems of stochastic chemical kinetics that is also based on Lyapunov theory is investigated. The technique explores states that are only in the set concentrating the steady-state probability mass and then resorts to polyhedra theory [46] to bound steady-state probabilities. Although the technique can work on states arbitrarily far away from the origin, it yields results that are much less accurate compared to those provided by the technique discussed in this chapter.

In closing, we point out the difficulties associated with analyzing systems of stochastic chemical kinetics that would be modeled as HMMs and SANs. Normally each countably infinite variable would be modeled as a separate subsystem with truncated state space. The control unit could still be modeled as a single subsystem with finite state space. Hence, it is possible to consider a model with $(H_I + 1)$ subsystems when $H_F > 0$ or H_I subsystems when $H_F = 0$. With SANs [151], there would be K Kronecker product terms of which those that have state-dependent $f^{(k,h)}(i_h)$ values for $h = 1, \ldots, H$ would be represented with functional transitions in the corresponding factors. For instance, this implies all transition classes in Table 5.1. Among these transition classes, the most difficult ones to work with would be those that have dependencies on subsystems associated with countably infinite variables (such as transition classes 2, 3, 6, 7, 10, 11 in Table 5.1). When executing the vector–Kronecker product multiplication algorithm, such transition classes would require a different evaluation for each value of the corresponding countably infinite variable, thereby complicating and slowing down the analysis process. As for HMMs, transition classes that have dependencies on subsystems associated with countably infinite variables would require a separate term for each state-dependent value of the variable (and as many terms as there are state-dependent values for the finite variable). Hence, each functional transition would need to be transformed into multiple Kronecker product terms. This not only would increase the number of Kronecker product terms by a number proportional to the sum of the truncated state space sizes of the countably infinite variables, but it would also make the corresponding factors in the Kronecker product terms extremely sparse, having a single nonzero entry. Therefore, any gain that would accrue from using a Kronecker-based representation would start to diminish.

Chapter 6
Conclusion

Transition matrices associated with MCs based on Kronecker products have a rich structure that is nested and recursive. Preprocessing techniques such as reordering, grouping, and lumping can take advantage of this structure to expedite analysis. All iterative analysis methods rely on an efficient vector–Kronecker product multiplication algorithm. Block iterative methods based on splittings, projection methods preconditioned with block iterative methods, and multilevel iterative methods for steady-state analysis come across as a strong set of solvers that should be integrated to software packages working with Kronecker products. Among these, multilevel methods perform better on a larger number of problems in the literature. Implementation of these solvers requires intricate programming with dynamically allocated, relatively complex data structures, which needs time, careful testing, and tuning. The major challenge in this process is to develop skills to be able to work with the abstract multidimensional state space utilizing Kronecker products, which can be visualized on paper in two dimensions but not so easily in higher dimensions.

To explain the introduced concepts further, Markovian models from availability, queueing networks, and stochastic chemical kinetics were considered. In each case, the transition matrix underlying the respective MC was expressed using Kronecker products. Approaches that can be employed to handle unreachable states and countably infinite state spaces were indicated. The advantages of using a decompositional iterative method for a particular kind of availability model to achieve fast convergence, two approximative decompositional iterative methods for closed queueing networks with phase-type service distributions and arbitrary buffer sizes when less accurate results can be tolerated, and a matrix-analytic method for systems of stochastic chemical kinetics having countably infinite state spaces were shown. Pointers to software used during this process were given. Efficient implementation of transient solvers and matrix-analytic solvers based on Kronecker products, especially for countably infinite state spaces, emerge as areas that require investigation.

T. Dayar, *Analyzing Markov Chains using Kronecker Products: Theory and Applications*, SpringerBriefs in Mathematics, DOI 10.1007/978-1-4614-4190-8_6,
© Tuğrul Dayar 2012

References

1. Akyildiz, I.F.: Mean value analysis for blocking queueing networks. IEEE Trans. Softw. Eng. **14**, 418–428 (1988)
2. Aldous, D., Shepp, L.: The least variable phase type distribution is Erlang. Stoch. Model. **3**, 467–473 (1987)
3. APNN–Toolbox. http://www4.cs.uni-dortmund.de/APNN-TOOLBOX/ (2004). Accessed 4 Apr 2012
4. Bao, Y., Bozkurt, I.N., Dayar, T., Sun, X., Trivedi, K.S.: Decompositional analysis of Kronecker structured Markov chains. Electron. Trans. Numer. Anal. **31**, 271–294 (2008)
5. Barker, V.A.: Numerical solution of sparse singular linear systems of equations arising from ergodic Markov chains. Comm. Stat. Stoch. Model. **5**, 355–381 (1989)
6. Baumann, H., Sandmann, W.: Numerical solution of level dependent quasi-birth-and-death processes. In: International Conference on Computational Science, Procedia Computer Science, vol. 1, pp. 1555–1563. Elsevier, Amsterdam (2010)
7. Bause, F., Buchholz, P., Kemper, P.: A toolbox for functional and quantitative analysis of DEDS. In: Puigjaner, R., Savino, N.N., Serra, B. (eds.) Quantitative Evaluation of Computing and Communication Systems, Lecture Notes in Computer Science, vol. 1469, pp. 356–359. Springer, Berlin Heidelberg New York (1998)
8. Benoit, A., Brenner, L., Fernandes, P., Plateau, B., Stewart, W.J.: The Peps software tool. In: Kemper, P., Sanders, W.H. (eds.) Computer Performance Evaluation: Modelling Techniques and Tools, Lecture Notes in Computer Science, vol. 2794, pp. 98–115. Springer, Heidelberg (2003)
9. Benoit, A., Brenner, L., Fernandes, P., Plateau, B.: Aggregation of stochastic automata networks with replicas. Linear Algebr. Appl. **386**, 111–136 (2004)
10. Benoit, A., Fernandes, P., Plateau, B., Stewart, W.J.: On the benefits of using functional transitions and Kronecker algebra. Perform. Eval. **58**, 367–390 (2004)
11. Benoit, A., Plateau, B., Stewart, W.J.: Memory-efficient Kronecker algorithms with applications to the modelling of parallel systems. Futur. Gener. Comput. Syst. **22**, 838–847 (2006)
12. Benzi, M: Preconditioning techniques for large linear systems: a survey. J. Comput. Phys. **182**, 418–477 (2002)
13. Berman, A., Plemmons, R.J.: Nonnegative Matrices in the Mathematical Sciences. SIAM, Philadelphia, Pennslyvania (1994)
14. Bini, D.A., Latouche, G., Meini, B.: Numerical Methods for Structured Markov Chains. Oxford University, Oxford (2005)
15. Brenner, L., Fernandes, P., Plateau, B., Sbeity, I.: PEPS 2007 – Stochastic automata networks software tool. In: Proceedings of the Fourth International Conference on Quantitative Evaluation of Computer Systems and Technologies, pp. 163–164. IEEE Computer Society, Edinburgh (2007)

T. Dayar, *Analyzing Markov Chains using Kronecker Products: Theory and*
Applications, SpringerBriefs in Mathematics, DOI 10.1007/978-1-4614-4190-8,
© Tuğrul Dayar 2012

16. Brenner, L., Fernandes, P., Fourneau, J.-M., Plateau, B.: Modelling Grid5000 point availability with SAN. Electron. Notes Theor. Comput. Sci. **232**, 165–178 (2009)

17. Briggs, W.L., Henson, V.E., McCormick, S.F.: A Multigrid Tutorial, 2nd edn. SIAM, Philadelphia, Pennslyvania (2000)

18. Bright, L., Taylor, P.G.: Calculating the equilibrium distribution in level dependent quasi-birth-and-death processes. Stoch. Model. **11**, 497–525 (1995)

19. Buchholz, P.: A class of hierarchical queueing networks and their analysis. Queue. Syst. **15**, 59–80 (1994)

20. Buchholz, P.: Exact and ordinary lumpability in finite Markov chains. J. Appl. Probab. **31**, 59–75 (1994)

21. Buchholz, P.: Hierarchical Markovian models: symmetries and reduction. Perform. Eval. **22**, 93–110 (1995)

22. Buchholz, P.: An aggregation\disaggregation algorithm for stochastic automata networks. Probab. Eng. Inf. Sci. **11**, 229–253 (1997)

23. Buchholz, P.: Exact performance equivalence: An equivalence relation for stochastic automata. Theor. Comput. Sci. **215**, 263–287 (1999)

24. Buchholz, P.: Hierarchical structuring of superposed GSPNs. IEEE Trans. Softw. Eng. **25**, 166–181 (1999)

25. Buchholz, P.: Structured analysis approaches for large Markov chains. Appl. Numer. Math. **31**, 375–404 (1999)

26. Buchholz, P.: Projection methods for the analysis of stochastic automata networks. In: Plateau, B., Stewart, W.J., Silva, M. (eds.) Numerical Solution of Markov Chains, pp. 149–168. Prensas Universitarias de Zaragoza, Zaragoza (1999)

27. Buchholz, P.: An adaptive aggregation/disaggregation algorithm for hierarchical Markovian models. Eur. J. Oper. Res. **116**, 545–564 (1999)

28. Buchholz, P.: Multilevel solutions for structured Markov chains. SIAM J. Matrix Anal. Appl. **22**, 342–357 (2000)

29. Buchholz, P.: Efficient computation of equivalent and reduced representations for stochastic automata. Comput. Syst. Sci. Eng. **15**, 93–103 (2000)

30. Buchholz, P.: An iterative bounding method for stochastic automata networks. Perform. Eval. **49**, 211–226 (2002)

31. Buchholz, P.: Adaptive decomposition and approximation for the analysis of stochastic Petri nets. Perform. Eval. **56**, 23–52 (2004)

32. Buchholz, P., Dayar, T.: Block SOR for Kronecker structured Markovian representations. Linear Algebr. Appl. **386**, 83–109 (2004)

33. Buchholz, P., Dayar, T.: Comparison of multilevel methods for Kronecker structured Markovian representations. Computing **73**, 349–371 (2004)

34. Buchholz, P., Dayar, T.: Block SOR preconditioned projection methods for Kronecker structured Markovian representations. SIAM J. Sci. Comput. **26**, 1289–1313 (2005)

35. Buchholz, P., Dayar, T.: On the convergence of a class of multilevel methods for large, sparse Markov chains. SIAM J. Matrix Anal. Appl. **29**, 1025–1049 (2007)

36. Buchholz, P., Kemper, P.: On generating a hierarchy for GSPN analysis. Perform. Eval. Rev. **26**, 5–14 (1998)

37. Buchholz, P., Kemper, P.: Kronecker based representations of large Markov chains. In: Haverkort, B., Hermanns, H., Siegle, M. (eds.) Validation of Stochastic Systems, Lecture Notes in Computer Science, vol. 2925, pp. 256–295. Springer, Berlin Heidelberg New York (2004)

38. Buchholz, P., Ciardo, G., Donatelli, S., Kemper, P.: Complexity of memory-efficient Kronecker operations with applications to the solution of Markov models. INFORMS J. Comput. **12**, 203–222 (2000)

39. Campos, J., Donatelli, S., Silva, M.: Structured solution of asynchronously communicating stochastic models. IEEE Trans. Softw. Eng. **25**, 147–165 (1999)

40. Cao, W.-L., Stewart, W.J.: Aggregation/disaggregation methods for nearly uncoupled Markov chains. J. ACM **32**, 702–719 (1985)

41. Chan, R.H., Ching, W.K.: Circulant preconditioners for stochastic automata networks. Numer. Math. **87**, 35–57 (2000)

42. Chung, M.-Y., Ciardo, G., Donatelli, S., He, N., Plateau, B., Stewart, W., Sulaiman, E., Yu, J.: A comparison of structural formalisms for modeling large Markov models. In: Proceedings of the 18th International Parallel and Distributed Processing Symposium, pp. 196b. IEEE Computer Society, Edinburgh (2004)

43. Ciardo, G., Miner, A.S.: A data structure for the efficient Kronecker solution of GSPNs. In: Buchholz, P., Silva, M. (eds.) Proceedings of the 8th International Workshop on Petri Nets and Performance Models, pp. 22–31. IEEE Computer Society, Edinburgh (1999)

44. Ciardo, G., Jones, R.L., Miner, A.S., Siminiceanu, R.: Logical and stochastic modeling with SMART. In: Kemper, P., Sanders, W.H. (eds.) Computer Performance Evaluation: Modelling Techniques and Tools, Lecture Notes in Computer Science, vol. 2794, pp. 78–97. Springer, Heidelberg (2003)

45. Clark, G., Gilmore, S., Hillston, J., Thomas, N.: Experiences with the PEPA performance modelling tools. IEE Softw. **146**, 11–19 (1999)

46. Courtois, P.-J., Semal, P.: Bounds for the positive eigenvectors of nonnegative matrices and for their approximations by decomposition. J. ACM **31**, 804-825 (1984)

47. Czekster, R.M., Fernandes, P., Vincent, J.-M., Webber, T.: Split: a flexible and efficient algorithm to vector–descriptor product. In: Glynn, P.W. (ed.) Proceedings of the 2nd International Conference on Performance Evaluation Methodologies and Tools, 83. Nantes, ACM International Conference Proceeding Series (2007)

48. Czekster, R.M., Fernandes, P., Webber, T.: GTAexpress: A software package to handle Kronecker descriptors. In: Proceedings of the Sixth International Conference on Quantitative Evaluation of Computer Systems and Technologies, pp. 281–282. IEEE Computer Society, Budapest (2009)

49. Dao-Thi, T.-H., Fourneau, J.-F.: Stochastic automata networks with master/slave synchronization: Product form and tensor. In: Al-Begain, K., Fiems, D., Horvaáthe, G. (eds.) Proceedings of the 16th International Conference on Analytical and Stochastic Modeling Techniques and Applications, Lecture Notes in Computer Science, vol. 5513, pp. 279–293. Springer, Heidelberg (2009)

50. Davio, M.: Kronecker products and shuffle algebra. IEEE Trans. Comput. **C-30**, 116–125 (1981)

51. Davis, T.A., Gilbert, J.R., Larimore, S., Ng, E.: Algorithm 836: COLAMD, a column approximate minimum degree ordering algorithm. ACM Trans. Math. Softw. **30**, 377–380 (2004)

52. Dayar, T.: State space orderings for Gauss–Seidel in Markov chains revisited. SIAM J. Sci. Comput. **19**, 148–154 (1998)

53. Dayar, T.: Permuting Markov chains to nearly completely decomposable form. Technical Report BU–CEIS–9808, Department of Computer Engineering and Information Science, Bilkent University, Ankara (1998)

54. Dayar, T.: Effects of reordering and lumping in the analysis of discrete–time SANs. In: Gardy, D., Mokkadem, A. (eds.) Mathematics and Computer Science: Algorithms, Trees, Combinatorics and Probabilities, pp. 209–220. Birkhauser, Switzerland (2000)

55. Dayar, T.: Analyzing Markov chains based on Kronecker products. In: Langville, A.N., Stewart, W.J. (eds.) MAM 2006: Markov Anniversary Meeting, pp. 279–300. Boson Books, Raleigh, North Carolina (2006)

56. Dayar, T., Meriç, A.: Kronecker representation and decompositional analysis of closed queueing networks with phase-type service distributions and arbitrary buffer sizes. Ann. Oper. Res. **164**, 193–210 (2008)

57. Dayar, T., Orhan, M.C.: LDQBD solver version 2. http://www.cs.bilkent.edu.tr/~tugrul/software.html (2011). Accessed 4 Apr 2012

58. Dayar, T., Stewart, W.J.: Quasi lumpability, lower-bounding coupling matrices, and nearly completely decomposable Markov chains. SIAM J. Matrix Anal. Appl. **18**, 482–498 (1997)

59. Dayar, T., Stewart, W.J.: Comparison of partitioning techniques for two-level iterative solvers on large, sparse Markov chains. SIAM J. Sci. Comput. **21**, 1691–1705 (2000)

60. Dayar, T., Pentakalos, O.I., Stephens, A.B.: Analytical modeling of robotic tape libraries using stochastic automata. Technical Report TR–97–189, Center of Excellence in Space Data & Information Systems, NASA/Goddard Space Flight Center, Greenbelt, Maryland (1997)

61. Dayar, T., Hermanns, H., Spieler, D., Wolf, V.: Bounding the equilibrium distribution of Markov population models. Numer. Linear Algebr. Appl. **18**, 931–946 (2011)

62. Dayar, T., Sandmann, W., Spieler, D., Wolf, V.: Infinite level–dependent QBDs and matrix analytic solutions for stochastic chemical kinetics. Adv. Appl. Probab. **43**, 1005–1026 (2011)

63. Donatelli, S.: Superposed stochastic automata: a class of stochastic Petri nets with parallel solution and distributed state space. Perform. Eval. **18**, 21–26 (1993)

64. Duff, I.S., Erisman, A.M., Reid, J.K.: Direct Methods for Sparse Matrices. Clarendon, Oxford (1986)

65. Fernandes, P., Plateau, B.: Triangular solution of linear systems in tensor product format. Perform. Eval. Rev. **28**(4), 30–32 (2001)

66. Fernandes, P., Plateau, B., Stewart, W.J.: Efficient descriptor–vector multiplications in stochastic automata networks. J. ACM **45**, 381–414 (1998)

67. Fernandes, F., Plateau, B., Stewart, W.J.: Optimizing tensor product computations in stochastic automata networks. RAIRO Oper. Res. **32**, 325–351 (1998)

68. Fletcher, R.: Conjugate gradient methods for indefinite systems. In: Watson, G.A. (ed.) Proceedings of the Dundee Conference on Numerical Analysis, Lecture Notes in Mathematics, vol. 506, pp. 73–89. Springer, Heidelberg (1976)

69. Fourneau, J.-M.: Discrete time stochastic automata networks: using structural properties and stochastic bounds to simplify the SAN. In: Glynn, P.W. (ed.) Proceedings of the 2nd International Conference on Performance Evaluation Methodologies and Tools, 84. Nantes, ACM International Conference Proceeding Series (2007)

70. Fourneau, J.-M.: Product form steady-state distribution for stochastic automata networks with domino synchronizations. In: Thomas, N., Juiz, C. (eds.) Proceedings of the 5th European Performance Engineering Workshop, Lecture Notes in Computer Science, vol. 5261, pp. 110–124. Springer, Berlin Heidelberg New York (2008)

71. Fourneau, J.-M.: Collaboration of discrete-time Markov chains: tensor and product form. Perform Eval. **67**, 779–796 (2010)

72. Fourneau, J.-M., Quessette, F.: Graphs and stochastic automata networks. In: Stewart, W.J. (ed.) Computations with Markov Chains. In: Proceedings of the 2nd International Workshop on the Numerical Solution of Markov Chains, pp. 217–235. Kluwer, Boston (1995)

73. Fourneau, J.-M., Maisonniaux, H., Pekergin, N., Véque, V.: Performance evaluation of a buffer policy with stochastic automata networks. In: IFIP Workshop on Modelling and Performance Evaluation of ATM Technology, vol. C–15, pp. 433–451. La Martinique, IFIP Transactions North-Holland, Amsterdam (1993)

74. Fourneau, J.-M., Kloul, L., Pekergin, N., Quessette, F., Véque, V.: Modelling buffer admission mechanisms using stochastic automata networks. Rev. Ann. Télécommun. **49**, 337–349 (1994)

75. Fourneau, J.-M., Plateau, B., Stewart, W.J.: An algebraic condition for product form in stochastic automata networks without synchronizations. Perform. Eval. **65**, 854–868 (2008)

76. Freund, R.W., Nachtigal, N.M.: QMR: a quasi-minimal residual method for non-Hermitian linear systems. Numer. Math. **60**, 315–339 (1991)

77. Gillespie, D.T.: Exact stochastic simulation of coupled chemical reactions. J. Phys. Chem. **81**, 2340–2361 (1977)

78. Golub, G.H., Van Loan, C.F.: Matrix Computations, 3rd edn. Johns Hopkins University, Baltimore (1996)

79. Gordon, J.W., Newell, G.F.: Closed queueing systems with exponential servers. Oper. Res. **15**, 252–267 (1967)

80. Grassmann, W.K.: Transient solutions in Markovian queueing systems. Comput. Oper. Res. **4**, 47–56 (1977)

81. Grassmann, W.K. (ed.): Computational Probability. Kluwer, Norwell, MA (2000)
82. Grassmann, W.K., Stanford, D.A.: Matrix analytic methods. In: Grassmann, W.K. (ed.) Computational Probability, pp. 153–204. Kluwer, Norwell, MA (2000)
83. Greenbaum, A.: Iterative Methods for Solving Linear Systems. SIAM, Philadelphia, Pennslyvania (1997)
84. Gross, D., Miller, D.R.: The randomization technique as a modeling tool and solution procedure for transient Markov processes. Oper. Res. **32**, 343–361 (1984)
85. Gusak, O., Dayar, T.: Iterative aggregation–disaggregation versus block Gauss–Seidel on continuous-time stochastic automata networks with unfavorable partitionings. In: Obaidat, M.S., Davoli, F. (eds.) Proceedings of the 2001 International Symposium on Performance Evaluation of Computer and Telecommunication Systems, pp. 617–623. Orlando, Florida (2001)
86. Gusak, O., Dayar, T., Fourneau, J.-M.: Stochastic automata networks and near complete decomposability. SIAM J. Matrix Anal. Appl. **23**, 581–599 (2001)
87. Gusak, O., Dayar, T., Fourneau, J.-M.: Lumpable continuous-time stochastic automata networks. Eur. J. Oper. Res. **148**, 436–451 (2003)
88. Gusak, O., Dayar, T., Fourneau, J.-M.: Iterative disaggregation for a class of lumpable discrete-time stochastic automata networks. Perform. Eval. **53**, 43–69 (2003)
89. Haddad, S., Moreaux, P.: Asynchronous composition of high–level Petri nets: a quantitative approach. In: Billington, J., Reisig, W. (eds.) Proceedings of the 17th International Conference on Application and Theory of Petri Nets, Lecture Notes in Computer Science, vol. 1091, pp. 192–211. Springer, Heidelberg (1996)
90. Haverkort, B.R.: Performance of Computer Communication Systems: A Model-Based Approach. Wiley, New York (1998)
91. Hillston, J., Kloul, L.: An efficient Kronecker representation for PEPA models. In: de Alfaro, L., Gilmore, S. (eds.) Proceedings of the 1st Process Algebras and Performance Modeling, Probabilistic Methods in Verification Workshop, Lecture Notes in Computer Science, vol. 2165, pp. 120–135. Springer, Berlin Heidelberg New York (2001)
92. Horton, G., Leutenegger, S.: A multi-level solution algorithm for steady state Markov chains. Perform. Eval. Rev. **22**(1), 191–200 (1994)
93. Kemeny, J.G., Snell, J.L.: Finite Markov Chains. Springer, Berlin Heidelberg New York (1983)
94. Kemper, P.: Numerical analysis of superposed GSPNs. IEEE Trans. Softw. Eng. **22**, 615–628 (1996)
95. Koury, J.R., McAllister, D.F., Stewart, W.J.: Iterative methods for computing stationary distributions of nearly completely decomposable Markov chains. SIAM J. Algebr. Discrete Math. **5**, 164–186 (1984)
96. Krieger, U.: Numerical solution of large finite Markov chains by algebraic multigrid techniques. In: Stewart, W.J. (ed.) Computations with Markov Chains, pp. 403–424. Kluwer, Boston (1995)
97. Kurtz, T.G.: The relationship between stochastic and deterministic models for chemical reactions. J. Chem. Phys. **57**, 2976–2978 (1972)
98. Langville, A.N., Stewart, W.J.: The Kronecker product and stochastic automata networks. J. Comput. Appl. Math. **167**, 429–447 (2004)
99. Langville, A.N., Stewart, W.J.: Testing the nearest Kronecker product preconditioner on Markov chains and stochastic automata networks. INFORMS J. Comput. **16**, 300–315 (2004)
100. Langville, A.N., Stewart, W.J.: A Kronecker product approximate preconditioner for SANs. Numer. Linear Algebr. Appl. **11**, 723–752 (2004)
101. Latouche, G., Ramaswami, V.: Introduction to Matrix Analytic Methods in Stochastic Modeling. SIAM, Philadelphia, Pennslyvania (1999)
102. Lee, T.L., Li, T.Y., Tsai, C.H.: HOM4PS-2.0: A software package for solving polynomial systems by the polyhedral homotopy continuation method. Computing **83**, 109–133 (2008)
103. Li, H., Cao, Y., Petzold, L.R., Gillespie, D.: Algorithms and software for stochastic simulation of biochemical reacting systems. Biotechnol. Prog. **24**, 56–62 (2008)

104. Loinger, A., Biham, O.: Stochastic simulations of the repressilator circuit. Phys. Rev. E **76**, 051917 (2007)

105. Marek, I., Mayer, P.: Convergence analysis of an iterative aggregation/disaggregation method for computing stationary probability vectors of stochastic matrices. Numer. Linear Algebr. Appl. **5**, 253–274 (1998)

106. Marek, I., Pultarová, I.: A note on local and global convergence analysis of iterative aggregation–disaggregation methods. Linear Algebra Appl. **413**, 327-341 (2006)

107. Marie, A.R.: An approximate analytical method for general queueing networks. IEEE Trans. Softw. Eng. **5**, 530–538 (1979)

108. Meriç, A.: Kronecker Representation and Decompositional Analysis of Closed Queueing Networks with Phase–Type Service Distributions and Arbitrary Buffer Sizes. M.S. Thesis, Department of Computer Engineering, Bilkent University, Ankara, Turkey (2007)

109. Meriç, A.: Software for Kronecker Representation and Decompositional Analysis of Closed Queueing Networks with Phase-Type Service Distributions and Arbitrary Buffer Sizes. http://www.cs.bilkent.edu.tr/~tugrul/software.html (2007). Accessed 4 Apr 2012

110. Meyer, C.D.: Stochastic complementation, uncoupling Markov chains, and the theory of nearly reducible systems. SIAM Rev. **31**, 240–272 (1989)

111. Meyer, C.D.: Matrix Analysis and Applied Linear Algebra. SIAM, Philadelphia (2000)

112. Migallón, V., Penadés, J., Syzld, D.B.: Block two-stage methods for singular systems and Markov chains. Numer. Linear Algebr. Appl. **3**, 413–426 (1996)

113. Neuts, M.F.: Matrix-Geometric Solutions in Stochastic Models: An Algorithhmic Approach. Johns Hopkins University Press, Baltimore (1981)

114. Neuts, M.F.: Structured Stochastic Matrices of M/G/1 Type and Their Applications. Marcel Dekker, New York (1989)

115. Orhan, M.C.: Kronecker-based Infinite Level-Dependent QBDs: Matrix Analytic Solution versus Simulation. M.S. Thesis, Department of Computer Engineering, Bilkent University, Ankara, Turkey (2011)

116. Paige, R., Tarjan, R.E.: Three partition refinement algorithms. SIAM J. Comput. **16**, 973–989 (1987)

117. PEPA Home Page. http://www.dcs.ed.ac.uk/pepa/tools/ (2005). Accessed 4 Apr 2012

118. PEPS Home Page. http://www-id.imag.fr/Logiciels/peps (2007). Accessed 4 Apr 2012

119. Plateau, B.: On the stochastic structure of parallelism and synchronization models for distributed algorithms. Perform. Eval. Rev. **13**(2), 147–154 (1985)

120. Plateau, B., Atif, K.: Stochastic automata network for modeling parallel systems. IEEE Trans. Softw. Eng. **17**, 1093–1108 (1991)

121. Plateau, B., Fourneau, J.-M.: A methodology for solving Markov models of parallel systems. J. Parallel Distrib. Comput. **12**, 370–387 (1991)

122. Plateau, B., Stewart, W.J.: Stochastic automata networks. In: W.K. Grassmann, W.K. (ed.) Computational Probability, pp. 113–152. Kluwer, Norwell, MA (2000)

123. Plateau, B.D., Tripathi, S.K.: Performance analysis of synchronization for two communicating processes. Perform. Eval. **8**, 305–320 (1988)

124. Plateau, B., Fourneau, J.-M., Lee, K.-H.: PEPS: A package for solving complex Markov models of parallel systems. In: Puigjaner, R., Ptier, D. (eds.) Modeling Techniques and Tools for Computer Performance Evaluation, pp. 291–305. Palma de Mallorca (1988)

125. Pultarová, I., Marek, I.: Convergence of multi-level iterative aggregation–disaggregation methods. J. Comp. Appl. Math **236**, 354–363 (2011)

126. Ramaswami, V., Taylor, P.G.: Some properties of the rate operators in level dependent quasi-birth-and-death processes with a countable number of phases. Stoch. Model. **12**, 143–164 (1996)

127. Ruge, J.W., Stüben, K.: Algebraic multigrid. In: McCormick, S.F. (ed.) Multigrid Methods, Frontiers in Applied Mathematics 3, pp. 73–130. SIAM, Philadelphia (1987)

128. Saad, Y.: Projection methods for the numerical solution of Markov chain models. In: Stewart, W.J. (ed.) Numerical Solution of Markov Chains, pp. 455–471. Marcel Dekker, New York (1991)

129. Saad, Y.: Preconditioned Krylov subspace methods for the numerical solution of Markov chains. In: Stewart, W.J. (ed.) Computations with Markov Chains. In: Proceedings of the 2nd International Workshop on the Numerical Solution of Markov Chains, pp. 49–64. Kluwer, Boston (1995)
130. Saad, Y.: Iterative Methods for Sparse Linear Systems. SIAM, Philadelphia (2003)
131. Saad, Y., Schultz, M.H.: GMRES: a generalized minimum residual algorithm for nonsymmetric linear systems. SIAM J. Sci. Stat. Comput. **7**, 856–869 (1986)
132. Sbeity, I., Plateau, B.: Structured stochastic modeling and performance analysis of a multiprocessor system. In: Langville, A.N., Stewart, W.J. (eds.) MAM 2006: Markov Anniversary Meeting, pp. 301–314. Boson Books, Raleigh, NC (2006)
133. Sbeity, I., Brenner, L., Plateau, B., Stewart, W.J.: Phase-type distributions in stochastic automata networks. Eur. J. Oper. Res. **186**, 1008–1028 (2008)
134. Scarpa, M., Bobbio, A.: Kronecker representation of stochastic Petri nets with discrete PH distributions. In: Proceedings of the IEEE International Computer Performance and Dependability Symposium, pp. 52–61. IEEE Computer Society, Budapest (1998)
135. Seneta E.: Non-negative Matrices: An Introduction to Theory and Applications. Allen & Unwin, London (1973)
136. SMART Project Home page. http://www.cs.ucr.edu/~ciardo/SMART (2004). Accessed 4 April 2012
137. Sonneveld, P.: CGS: A fast Lanczos-type solver for nonsymmetric linear systems. SIAM J. Sci. Stat. Comput. **10**, 36–52 (1989)
138. Stewart, G.W., Stewart, W.J., McAllister, D.F.: A two-stage iteration for solving nearly completely decomposable Markov chains. In: Golub, G.H., Greenbaum, A., Luskin, M. (eds.) The IMA Volumes in Mathematics and its Applications 60: Recent Advances in Iterative Methods, pp. 201–216. Springer, Berlin Heidelberg New York (1994)
139. Stewart, W.J.: Introduction to the Numerical Solution of Markov Chains. Princeton University Press, Princeton, NJ (1994)
140. Stewart, W.J.: Probability, Markov Chains, Queues, and Simulation: The Mathematical Basis of Performance Modeling. Princeton University Press, Princeton, NJ (2009)
141. Stewart, W.J., Atif, K., Plateau, B.: The numerical solution of stochastic automata networks. Eur. J. Oper. Res. **86**, 503–525 (1995)
142. StochKit. http://engineering.ucsb.edu/~cse/StochKit/ (2012). Accessed 4 Apr 2012
143. Tewarson, R.P.: Sparse Matrices. Academic, New York (1973)
144. Touzene, A.: A tensor sum preconditioner for stochastic automata networks. INFORMS J. Comput. **20**, 234–242 (2008)
145. Tweedie, R.L.: Sufficient conditions for regularity, recurrence and ergodicity of Markov processes. Math. Proc. Camb. Philos. Soc. **78**, 125–136 (1975)
146. Uysal, E., Dayar, T.: Iterative methods based on splittings for stochastic automata networks. Eur. J. Oper. Res. **110**, 166–186 (1998)
147. van der Vorst, H.A.: BI-CGSTAB: a fast and smoothly converging variant of BI-CG for the solution of nonsymmetric linear systems. SIAM J. Sci. Stat. Comput. **13**, 631–644 (1992)
148. Van Loan, C.F.: The ubiquitous Kronecker product. J. Comput. Appl. Math. **123**, 85–100 (2000)
149. Vèque, V., Ben–Othman, J.: MRAP: A multiservices resource allocation policy for wireless ATM network. Comput. Netw. ISDN Syst. **29**, 2187–2200 (1998)
150. Wesseling, P.: An Introduction to Multigrid Methods. Wiley, Chichester (1992)
151. Wolf, V.: Modelling of biochemical reactions by stochastic automata networks. Electron. Notes Theor. Comput. Sci. **171**, 197–208 (2007)
152. Yao, D.D., Buzacott, J.A.: The exponentialization approach to flexible manufacturing systems models with general processing times. Eur. J. Oper. Res. **24**, 410–416 (1986)

Index